SUMMARY OF A WORKSHOP ON

INFORMATION
TECHNOLOGY
RESEARCH
for
Federal Statistics

Committee on Computing and Communications Research to Enable Better
Use of Information Technology in Government

Computer Science and Telecommunications Board
Commission on Physical Sciences, Mathematics, and Applications

Committee on National Statistics
Commission on Behavioral and Social Sciences and Education

National Research Council

NATIONAL ACADEMY PRESS
Washington, D.C.

NOTICE: The project that is the subject of this report was approved by the Governing Board of the National Research Council, whose members are drawn from the councils of the National Academy of Sciences, the National Academy of Engineering, and the Institute of Medicine. The members of the committee responsible for the report were chosen for their special competences and with regard for appropriate balance.

Support for this project was provided by the National Science Foundation under grant EIA-9809120. Support for the work of the Committee on National Statistics is provided by a consortium of federal agencies through a grant between the National Academy of Sciences and the National Science Foundation (grant number SBR-9709489). Any opinions, findings, conclusions, or recommendations expressed in this material are those of the authors and do not necessarily reflect the views of the sponsor.

International Standard Book Number 0-309-07097-X

Additional copies of this report are available from:

National Academy Press (http://www.nap.edu)
2101 Constitution Ave., NW, Box 285
Washington, D.C. 20055
800-624-6242
202-334-3313 (in the Washington metropolitan area)

Copyright 2000 by the National Academy of Sciences. All rights reserved.

Printed in the United States of America

THE NATIONAL ACADEMIES

National Academy of Sciences
National Academy of Engineering
Institute of Medicine
National Research Council

The **National Academy of Sciences** is a private, nonprofit, self-perpetuating society of distinguished scholars engaged in scientific and engineering research, dedicated to the furtherance of science and technology and to their use for the general welfare. Upon the authority of the charter granted to it by the Congress in 1863, the Academy has a mandate that requires it to advise the federal government on scientific and technical matters. Dr. Bruce M. Alberts is president of the National Academy of Sciences.

The **National Academy of Engineering** was established in 1964, under the charter of the National Academy of Sciences, as a parallel organization of outstanding engineers. It is autonomous in its administration and in the selection of its members, sharing with the National Academy of Sciences the responsibility for advising the federal government. The National Academy of Engineering also sponsors engineering programs aimed at meeting national needs, encourages education and research, and recognizes the superior achievements of engineers. Dr. William A. Wulf is president of the National Academy of Engineering.

The **Institute of Medicine** was established in 1970 by the National Academy of Sciences to secure the services of eminent members of appropriate professions in the examination of policy matters pertaining to the health of the public. The Institute acts under the responsibility given to the National Academy of Sciences by its congressional charter to be an adviser to the federal government and, upon its own initiative, to identify issues of medical care, research, and education. Dr. Kenneth I. Shine is president of the Institute of Medicine.

The **National Research Council** was organized by the National Academy of Sciences in 1916 to associate the broad community of science and technology with the Academy's purposes of furthering knowledge and advising the federal government. Functioning in accordance with general policies determined by the Academy, the Council has become the principal operating agency of both the National Academy of Sciences and the National Academy of Engineering in providing services to the government, the public, and the scientific and engineering communities. The Council is administered jointly by both Academies and the Institute of Medicine. Dr. Bruce M. Alberts and Dr. William A. Wulf are chairman and vice chairman, respectively, of the National Research Council.

COMMITTEE ON COMPUTING AND COMMUNICATIONS RESEARCH TO ENABLE BETTER USE OF INFORMATION TECHNOLOGY IN GOVERNMENT

WILLIAM SCHERLIS, Carnegie Mellon University, *Chair*
W. BRUCE CROFT, University of Massachusetts at Amherst
DAVID DeWITT, University of Wisconsin at Madison
SUSAN DUMAIS, Microsoft Research
WILLIAM EDDY, Carnegie Mellon University
EVE GRUNTFEST, University of Colorado at Colorado Springs
DAVID KEHRLEIN, Governor's Office of Emergency Services, State of California
SALLIE KELLER-McNULTY, Los Alamos National Laboratory
MICHAEL R. NELSON, IBM Corporation
CLIFFORD NEUMAN, Information Sciences Institute, University of Southern California

Staff

JON EISENBERG, Program Officer and Study Director
RITA GASKINS, Project Assistant (through September 1999)
DANIEL D. LLATA, Senior Project Assistant

COMPUTER SCIENCE AND TELECOMMUNICATIONS BOARD

DAVID D. CLARK, Massachusetts Institute of Technology, *Chair*
JAMES CHIDDIX, Time Warner Cable
JOHN M. CIOFFI, Stanford University
ELAINE COHEN, University of Utah
W. BRUCE CROFT, University of Massachusetts, Amherst
A.G. FRASER, AT&T Corporation
SUSAN L. GRAHAM, University of California at Berkeley
JUDITH HEMPEL, University of California at San Francisco
JEFFREY M. JAFFE, IBM Corporation
ANNA KARLIN, University of Washington
BUTLER W. LAMPSON, Microsoft Corporation
EDWARD D. LAZOWSKA, University of Washington
DAVID LIDDLE, Interval Research
TOM M. MITCHELL, Carnegie Mellon University
DONALD NORMAN, UNext.com
RAYMOND OZZIE, Groove Networks
DAVID A. PATTERSON, University of California at Berkeley
CHARLES SIMONYI, Microsoft Corporation
BURTON SMITH, Tera Computer Company
TERRY SMITH, University of California at Santa Barbara
LEE SPROULL, New York University

MARJORY S. BLUMENTHAL, Director
HERBERT S. LIN, Senior Scientist
JERRY R. SHEEHAN, Senior Program Officer
ALAN S. INOUYE, Program Officer
JON EISENBERG, Program Officer
GAIL PRITCHARD, Program Officer
JANET BRISCOE, Office Manager
DAVID DRAKE, Project Assistant
MARGARET MARSH, Project Assistant
DAVID PADGHAM, Project Assistant
MICKELLE RODGERS RODRIGUEZ, Senior Project Assistant
SUZANNE OSSA, Senior Project Assistant
DANIEL D. LLATA, Senior Project Assistant

COMMISSION ON PHYSICAL SCIENCES, MATHEMATICS, AND APPLICATIONS

PETER M. BANKS, Veridian ERIM International, Inc., *Co-chair*
W. CARL LINEBERGER, University of Colorado, *Co-chair*
WILLIAM F. BALLHAUS, JR., Lockheed Martin Corporation
SHIRLEY CHIANG, University of California at Davis
MARSHALL H. COHEN, California Institute of Technology
RONALD G. DOUGLAS, Texas A&M University
SAMUEL H. FULLER, Analog Devices, Inc.
JERRY P. GOLLUB, Haverford College
MICHAEL F. GOODCHILD, University of California at Santa Barbara
MARTHA P. HAYNES, Cornell University
WESLEY T. HUNTRESS, JR., Carnegie Institution
CAROL M. JANTZEN, Westinghouse Savannah River Company
PAUL G. KAMINSKI, Technovation, Inc.
KENNETH H. KELLER, University of Minnesota
JOHN R. KREICK, Sanders, a Lockheed Martin Company (retired)
MARSHA I. LESTER, University of Pennsylvania
DUSA M. McDUFF, State University of New York at Stony Brook
JANET L. NORWOOD, Former Commissioner, U.S. Bureau of Labor Statistics
M. ELISABETH PATÉ-CORNELL, Stanford University
NICHOLAS P. SAMIOS, Brookhaven National Laboratory
ROBERT J. SPINRAD, Xerox PARC (retired)

MYRON F. UMAN, Acting Executive Director

COMMITTEE ON NATIONAL STATISTICS

JOHN E. ROLPH, University of Southern California, *Chair*
JOSEPH G. ALTONJI, Northwestern University
LAWRENCE D. BROWN, University of Pennsylvania
JULIE DAVANZO, RAND, Santa Monica, California
WILLIAM F. EDDY, Carnegie Mellon University
HERMANN HABERMANN, United Nations, New York
WILLIAM D. KALSBEEK, University of North Carolina
RODERICK J.A. LITTLE, University of Michigan
THOMAS A. LOUIS, University of Minnesota
CHARLES F. MANSKI, Northwestern University
EDWARD B. PERRIN, University of Washington
FRANCISCO J. SAMANIEGO, University of California at Davis
RICHARD L. SCHMALENSEE, Massachusetts Institute of Technology
MATTHEW D. SHAPIRO, University of Michigan

ANDREW A. WHITE, Director

COMMISSION ON BEHAVIORAL AND SOCIAL SCIENCES AND EDUCATION

NEIL J. SMELSER, Center for Advanced Study in the Behavioral Sciences, Stanford, *Chair*
ALFRED BLUMSTEIN, Carnegie Mellon University
JACQUELYNNE ECCLES, University of Michigan
STEPHEN E. FIENBERG, Carnegie Mellon University
BARUCH FISCHHOFF, Carnegie Mellon University
JOHN F. GEWEKE, University of Iowa
ELEANOR E. MACCOBY, Stanford University
CORA B. MARRETT, University of Massachusetts
BARBARA J. McNEIL, Harvard Medical School
ROBERT A. MOFFITT, Johns Hopkins University
RICHARD J. MURNANE, Harvard University
T. PAUL SCHULTZ, Yale University
KENNETH A. SHEPSLE, Harvard University
RICHARD M. SHIFFRIN, Indiana University
BURTON H. SINGER, Princeton University
CATHERINE E. SNOW, Harvard University
MARTA TIENDA, Princeton University

BARBARA TORREY, Executive Director

Preface

As part of its new Digital Government program, the National Science Foundation (NSF) requested that the Computer Science and Telecommunications Board (CSTB) undertake an in-depth study of how information technology research and development could more effectively support advances in the use of information technology (IT) in government. CSTB's Committee on Computing and Communications Research to Enable Better Use of Information Technology in Government was established to organize two specific application-area workshops and conduct a broader study, drawing in part on those workshops, of how IT research can enable improved and new government services, operations, and interactions with citizens.

The committee was asked to identify ways to foster interaction among computing and communications researchers, federal managers, and professionals in specific domains that could lead to collaborative research efforts. By establishing research links between these communities and creating collaborative mechanisms aimed at meeting relevant requirements, NSF hopes to stimulate thinking in the computing and communications research community and throughout government about possibilities for advances in technology that will support a variety of digital initiatives by the government.

The first phase of the project focused on two illustrative application areas that are inherently governmental in nature—crisis management and federal statistics. In each of these areas, the study committee convened a workshop designed to facilitate interaction between stakeholders from the individual domains and researchers in computing and communications systems and to explore research topics that might be of relevance government-wide. The first workshop in the series explored information technology research for crisis management.[1] The second workshop, called "Information Technology Research for Federal Statistics" and held on February 9 and 10, 1999, in Washington, D.C., is summarized in this report.

Participants in the second workshop, which explored IT research opportunities of relevance to the collection, analysis, and dissemination of federal statistics, were drawn from a number of communities: IT research, IT research management, federal statistics, and academic statistics (see the appendix for the full agenda of the workshop and a list of participants). The workshop provided an opportunity for these communities to interact and to learn how they might collaborate more effectively in developing improved systems to support federal statistics. Two keynote speeches provided a foundation by describing developments in the statistics and information technology research communities. The first panel presented four case studies. Other panels then explored a range of ways in which IT is currently used in the federal statistical enterprise and articulated a set of challenges and opportunities for IT research in the collection, analysis, and dissemination of federal statistics. At the conclusion of the workshop, a set of parallel breakout sessions was held to permit workshop participants to look into opportunities for collaborative research between the IT and statistics communities and to identify some important research topics. This report is based on those presentations and discussions.

Because the development of specific requirements would of course be beyond the scope of a single workshop, this report cannot presume to be a comprehensive analysis of IT requirements in the federal statistical system. Nor does the report explore all aspects of the work of the federal statistical community. For example, the workshop did not specifically address the decennial census. Presentations and discussions focused on individual or household surveys; other surveys depend on data obtained from business and other organizations where there would, for example, be less emphasis on developing better survey interview instruments because the information is in many cases already being collected through automated systems. Because the workshop emphasized survey work in the federal statistical system, the report does not specifically address the full range of statistics applications that arise in the work of the federal government (e.g., biostatistical

---

[1]Computer Science and Telecommunications Board, National Research Council. 1999. *Summary of a Workshop on Information Technology Research for Crisis Management*. National Academy Press, Washington, D.C.

PREFACE   xi

work at the National Institutes of Health). However, by examining a representative range of IT applications, and through discussions between IT researchers and statistics professionals, the workshop was able to identify key issues that arise in the application of IT to federal statistics work and to explore possible research opportunities.

This report is an overview by the committee of topics covered and issues raised at the workshop. Where possible, related issues raised at various points during the workshop have been consolidated. In preparing the report, the committee drew on the contributions of speakers, panelists, and participants, who together richly illustrated the role of IT in federal statistics, issues surrounding its use, possible research opportunities, and process and implementation issues related to such research. To these contributions the committee added some context-setting material and examples. The report remains, however, primarily an account of the presentations and discussions at the workshop. Synthesis of the workshop experience into a more general, broader set of findings and recommendations for IT research in the digital government context was deferred to the second phase of the committee's work. This second phase is drawing on information from the two workshops, as well as from additional briefings and other work on the topic of digital government, to develop a final report that will provide recommendations for refining the NSF's Digital Government program and stimulating IT innovation more broadly across government.

Support for this project came from NSF, and the committee acknowledges Larry Brandt of the NSF for his encouragement of this effort. The National Research Council's Committee on National Statistics, CNSTAT, was a cosponsor of this workshop and provided additional resources in support of the project. This is a reporting of workshop discussions, and the committee thanks all participants for the insights they contributed through their workshop presentations, discussions, breakout sessions, and subsequent interactions. The committee also wishes to thank the CSTB staff for their assistance with the workshop and the preparation of the report. In particular, the committee thanks Jon Eisenberg, CSTB program officer, who made significant contributions to the organization of the workshop and the assembly of the report, which could not have been written without his help and facilitation. Jane Bortnick Griffith played a key role during her term as interim CSTB director in helping conceive and initiate this project. In addition, the committee thanks Daniel Llata for his contributions in preparing the report for publication. The committee also thanks Andy White from the National Research Council's Commission on Behavioral and Social Sciences and Education for his support and assistance with this project. Finally, the committee is grateful to the reviewers for helping to sharpen and improve the report through their comments. Responsibility for the report remains with the committee.

Acknowledgment of Reviewers

This report was reviewed by individuals chosen for their diverse perspectives and technical expertise, in accordance with the procedures approved by the National Research Council's (NRC's) Report Review Committee. The purpose of this independent review is to provide candid and critical comments that will assist the authors and the NRC in making the published report as sound as possible and to ensure that the report meets institutional standards for objectivity, evidence, and responsiveness to the study charge. The contents of the review comments and draft manuscript remain confidential to protect the integrity of the deliberative process. We wish to thank the following individuals for their participation in the review of this report:

Larry Brown, University of Pennsylvania,
Terrence Ireland, Consultant,
Diane Lambert, Bell Laboratories, Lucent Technologies,
Judith Lessler, Research Triangle Institute,
Teresa Lunt, SRI International,
Janet Norwood, Former Commissioner, U.S. Bureau of Labor Statistics,
Bruce Trumbo, California State University at Hayward, and
Ben Schneiderman, University of Maryland.

Although the individuals listed above provided many constructive comments and suggestions, responsibility for the final content of this report rests solely with the study committee and the NRC.

Contents

1 INTRODUCTION AND CONTEXT 1
  Overview of Federal Statistics, 1
  Activities of the Federal Statistics Agencies, 2
    Data Collection, 3
    Processing and Analysis, 7
    Creation and Dissemination of Statistical Products, 9
  Organization of the Federal Statistical System, 10
  Information Technology Innovation in Federal Statistics, 14

2 RESEARCH OPPORTUNITIES 17
  Human-Computer Interaction, 17
    User Focus, 19
    Universal Access, 19
    Literacy, Visualization, and Perception, 20
  Database Systems, 23
  Data Mining, 25
  Metadata, 29
  Information Integration, 30
  Survey Instruments, 31
  Limiting Disclosure, 34
  Trustworthiness of Information Systems, 41

3 INTERACTIONS FOR INFORMATION TECHNOLOGY
   INNOVATION IN FEDERAL STATISTICAL WORK   44

APPENDIX
   WORKSHOP AGENDA AND PARTICIPANTS   49

1

Introduction and Context

OVERVIEW OF FEDERAL STATISTICS

Federal statistics play a key role in a wide range of policy, business, and individual decisions that are made based on statistics produced about population characteristics, the economy, health, education, crime, and other factors. The decennial census population counts—along with related estimates that are produced during the intervening years—will drive the allocation of roughly $180 billion in federal funding annually to state and local governments.[1] These counts also drive the apportionment of legislative districts at the local, state, and federal levels. Another statistic, the Consumer Price Index, is used to adjust wages, retirement benefits, and other spending, both public and private. Federal statistical data also provide insight into the status, well-being, and activities of the U.S. population, including its health, the incidence of crime, unemployment and other dimensions of the labor force, and the nature of long-distance travel. The surveys conducted to derive this information (see the next section for examples) are extensive undertakings that involve the collection of detailed information, often from large numbers of respondents.

The federal statistical system involves about 70 government agencies. Most executive branch departments are, in one way or another, involved

---

[1] U.S. Census Bureau estimate from U.S. Census Bureau, Department of Commerce. 1999. *United States Census 2000: Frequently Asked Questions.* U.S. Census Bureau, Washington, D.C. Available online at <http://www.census.gov/dmd/www/faqquest.htm>.

in gathering and disseminating statistical information. The two largest statistical agencies are the Bureau of the Census (in the Department of Commerce) and the Bureau of Labor Statistics (in the Department of Labor). About a dozen agencies have statistics as their principal line of work, while others collect statistics in conjunction with other activities, such as administering a program benefit (e.g., the Health Care Financing Administration or the Social Security Administration) or promulgating regulations in a particular area (e.g., the Environmental Protection Agency). The budgets for all of these activities—excluding the estimated $6.8 billion cost of the decennial census[2]—total more than $3 billion per year.[3]

These federal statistical agencies are characterized not only by their mission of collecting statistical information but also by their independence and commitment to a set of principles and practices aimed at ensuring the quality and credibility of the statistical information they provide (Box 1.1). Thus, the agencies aim to live up to citizens' expectations for trustworthiness, so that citizens will continue to participate in statistical surveys, and to the expectations of decision makers, who rely on the integrity of the statistical products they use in policy formulation.

ACTIVITIES OF THE FEDERAL STATISTICS AGENCIES

Many activities take place in connection with the development of federal statistics—the planning and design of surveys (see Box 1.2 for examples of such surveys); data collection, processing, and analysis; and the dissemination of results in a variety of forms to a range of users. What follows is not intended as a comprehensive discussion of the tasks involved in creating statistical products; rather, it is provided as an outline of the types of tasks that must be performed in the course of a federal statistical survey. Because the report as a whole focuses on information technology (IT) research opportunities, this section emphasizes the IT-related aspects of these activities and provides pointers to pertinent discussions of research opportunities in Chapter 2.

---

[2]Estimate by Census Bureau director of total costs in D'Vera Cohn. 2000. "Early Signs of Census Avoidance," *Washington Post,* April 2, p. A8.

[3]For more details on federal statistical programs, see Executive Office of the President, Office of Management and Budget (OMB). 1998. *Statistical Programs of the United States Government.* OMB, Washington, D.C.

> **BOX 1.1**
> **Principles and Practices for a Federal Statistical Agency**
>
> In response to requests for advice on what constitutes an effective federal statistical agency, the National Research Council's Committee on National Statistics issued a white paper that identified the following as principles and best practices for federal statistical agencies:
>
> *Principles*
> - Relevance to policy issues
> - Credibility among data users
> - Trust among data providers and data subjects
>
> *Practices*
> - A clearly defined and well-accepted mission
> - A strong measure of independence
> - Fair treatment of data providers
> - Cooperation with data users
> - Openness about the data provided
> - Commitment to quality and professional standards
> - Wide dissemination of data
> - An active research program
> - Professional advancement of staff
> - Caution in conducting nonstatistical activities
> - Coordination with other statistical agencies
>
> SOURCE: Adapted from Margaret E. Martin and Miron L. Straf, eds. 1992. *Principles and Practices for a Federal Statistical Agency.* Committee on National Statistics, National Research Council. National Academy Press, Washington, D.C.

Data Collection

Data collection starts with the process of selection.[4] Ensuring that survey samples are representative of the populations they measure is a significant undertaking. This task entails first defining the population of interest (e.g., the U.S. civilian noninstitutionalized population, in the case of the National Health and Nutrition Examination Survey). Second, a

---

[4]This discussion focuses on the process of conducting surveys of individuals. Many surveys gather information from businesses or other organizations. In some instances, similar interview methods are used; in others, especially with larger organizations, the data are collected through automated processes that employ standardized reporting formats.

> **BOX 1.2**
> **Examples of Federal Statistical Surveys**
>
> To give workshop participants a sense of the range of activities and purposes of federal statistical surveys, representatives of several large surveys sponsored by federal statistical agencies were invited to present case studies at the workshop. Reference is made to several of these examples in the body of this report.
>
> *National Health and Nutrition Examination Survey*
>
> The National Health and Nutrition Examination Survey (NHANES) is one of several major data collection studies sponsored by the National Center for Health Statistics (NCHS). Under the legislative authority of the Public Health Service, NCHS collects statistics on the nature of illness and disability in the population; on environmental, nutritional, and other health hazards; and on health resources and utilization of health care. NHANES has been conducted since the early 1960s; its ninth survey is NHANES 1999.[1] It is now implemented as a continuous, annual survey in which a sample of approximately 5,000 individuals representative of the U.S. population is examined each year. Participants in the survey undergo a detailed home interview and a physical examination and health and dietary interviews in mobile examination centers set up for the survey. Home examinations, which include a subset of the exam components conducted at the exam center, are offered to persons unable or unwilling to come to the center for the full examination.
>
> The main objectives of NHANES are to estimate the prevalence of diseases and risks factors and monitoring trends for them; to explore emerging public health issues, such as cardiovascular disease; to correlate findings of health measures in the survey, such as body measurements and blood characteristics, and to establish a national probability sample of DNA materials using NHANES-collected blood samples. There are a variety of consumers for the NHANES data, including government agencies, state and local communities, private researchers, and companies, including health care providers. Findings from NHANES are used as the basis for such things as the familiar growth charts for children and material on obesity in the United States. For example, the body mass index used in understanding obesity is derived from NHANES data and was developed by the National Institutes of Health in collaboration with NCHS. Other findings, such as the effects of lead in gasoline and in paint and the effects of removing it, are also based on NHANES data.[2]
>
> ---
>
> [1] Earlier incarnations of the NHANES survey were called, first, the Health Examination Survey and then, the Health and Nutrition Examination Survey (HANES). Unlike previous surveys, NHANES 1999 is intended to be a continuous survey with ongoing data collection.
> [2] This description is adapted in part from documents on the National Health and Nutrition Examination Survey Web site. (Department of Health and Human Services, Centers for Disease Control, National Center for Health Statistics (NCHS). 1999. National Health and Nutrition Examination Survey. Available online at <http://www.cdc.gov/nchswww/about/major/nhanes/nhanes.htm>.)
>
> *continued*

BOX 1.2 Continued

American Travel Survey

The American Travel Survey (ATS), sponsored by the Department of Transportation, tracks passenger travel throughout the United States. The first primary objective is to obtain information about long-distance travel [3] by persons living in the United States. The second primary objective is to inform policy makers about the principal characteristics of travel and travelers, such as the frequency and economic implications of long-distance travel, which are useful for a variety of planning purposes. ATS is designed to provide reliable estimates at national and state levels for all persons and households in the United States—frequency, primary destinations, mode of travel (car, plane, bus, train, etc.), and purpose. Among the other data collected by the ATS is the flow of travel between states and between metropolitan areas.

The survey samples approximately 80,000 households in the United States and conducts interviews with about 65,000 of them, making it the second largest (after the decennial census) household survey conducted by federal statistical agencies. Each household is interviewed four times in a calendar year to yield a record of the entire year's worth of long-distance travel; in each interview, a household is asked to recall travel that occurred in the preceding 3 months. Information is collected by computer-assisted telephone interviewing (CATI) systems as well as via computer-assisted personal interviewing (CAPI).

Current Population Survey

The primary goal of the Current Population Survey (CPS), sponsored by the Bureau of Labor Statistics (BLS), is to measure the labor force. Collecting demographic and labor force information on the U.S. population age 16 and older, the CPS is the source of the unemployment numbers reported by BLS on the first Friday of every month. Initiated more than 50 years ago, it is the longest-running continuous monthly survey in the United States using a statistical sample. Conducted by the Census Bureau for BLS, the CPS is the largest of the Census Bureau's ongoing monthly surveys. It surveys about 50,000 households; the sample is divided into eight representative subsamples. Each subsample group is interviewed for a total of 8 months—in the sample for 4 consecutive months, out of the sample during the following 8 months, and then back in the sample for another 4 consecutive months. To provide better estimates of change and reduce discontinuities without overly burdening households with a long period of participation, the survey is conducted on a rotating basis so that 75 percent of the sample is common from month to month and 50 percent from year to year for the same month.[4]

---

[3] Long-distance is defined in the ATS as a trip of 100 miles or more. The Nationwide Personal Transportation Survey (NPTS) collects data on daily, local passenger travel, covering all types and modes of trips. For further information, see the Bureau of Transportation's Web page on the NPTS, available online at <http://www.nptsats2000.bts.gov/>.

[4] For more details on the sampling procedure, see, for example the U.S. Census Bureau. 1997. *CPS Basic Monthly Survey: Sampling.* U.S. Census Bureau, Washington, D.C. Available online at <http://www.bls.census.gov/cps/bsampdes.htm>.

continued

> **BOX 1.2 Continued**
>
> Since the survey is designed to be representative of the U.S. population, a considerable quantity of useful information about the demographics of the U.S. population other than labor force data can be obtained from it, including occupations and the industries in which workers are employed. An important attribute of the CPS is that, owing to the short time required to gather the basic labor force information, the survey can easily be supplemented with additional questions. For example, every March, a supplement collects detailed income and work experience data, and every other February information is collected on displaced workers. Other supplements are conducted for a variety of agencies, including the Department of Veterans Affairs and the Department of Education.
>
> *National Crime Victimization Survey*
>
> The National Crime Victimization Survey (NCVS), sponsored by the Bureau of Justice Statistics, is a household-based survey that collects data on the amount and types of crime in the United States. Each year, the survey obtains data from a nationally representative sample of approximately 43,000 households (roughly 80,000 persons). It measures the incidence of violence against individuals, including rape, robbery, aggravated assault and simple assault, and theft directed at individuals and households, including burglary, motor vehicle theft, and household larceny. Other types of crimes, such as murder, kidnapping, drug abuse, prostitution, fraud, commercial burglary, and arson, are outside the scope of the survey. The NCVS, initiated in 1972, is one of two Department of Justice measures of crime in the United States, and it is intended to complement what is known about crime from the Federal Bureau of Investigation's annual compilation of information reported to law enforcement agencies (the Uniform Crime Reports). The NCVS serves two broad goals. First, it provides a time series tracing changes in both the incidence of crime and the various factors associated with criminal victimization. Second, it provides data that can be used to study particular research questions related to criminal victimization, including the relationship of victims to offenders and the costs of crime. Based on the survey, the Bureau of Justice Statistics publishes annual estimates of the national crime rate.[5]
>
> ---
>
> [5]Description adapted in part from U.S. Department of Justice, Bureau of Justice Statistics (BJS). 1999. *Crime and Victims Statistics*. BJS, Washington, D.C. Available online at <http://www.ojp.usdoj.gov/bjs/cvict.htm#ncvs>.

listing, or sample frame, is constructed. Third, a sample of appropriate size is selected from the sampling frame. There are many challenges associated with the construction of a truly representative sample: a sample frame of all households may require the identification of all housing units that have been constructed since the last decennial census was conducted. Also, when a survey is to be representative of a subpopulation (e.g., when the sample must include a certain number of children between the ages of 12 and 17), field workers may need to interview households or individuals to select appropriate participants.

Once a set of individuals or households has been identified for a survey, their participation must be tracked and managed, including assignment of individuals or households to interviewers, scheduling of telephone interviews, and follow-up with nonrespondents. A variety of techniques, generally computer-based, are used to assist field workers in conducting interviews (Box 1.3). Finally, data from interviews are collected from individual field interviewers and field offices for processing and analysis. Data collected from paper-and-pencil interviews, of course, require data entry (keying) prior to further processing.[5]

Processing and Analysis

Before they are included in the survey data set, data from respondents are subject to editing. Responses are checked for missing items and for internal consistency; cases that fail these checks can be referred back to the interviewer or field office for correction. The timely transmission of data to a location where such quality control measures can be performed allows rapid feedback to the field and increases the likelihood that corrected data can be obtained. In addition, some responses require coding before further processing. For example, in the Current Population Survey, verbal descriptions of industry and occupation are translated into a standardized set of codes. A variety of statistical adjustments, including a statistical procedure known as weighting, may be applied to the data to correct for errors in the sampling process or to impute nonresponses.

A wide variety of data-processing activities take place before statistical information products can be made available to the public. These activities depend on database systems; relevant trends in database technologies and research are discussed in the Chapter 2 section "Database Systems." In addition, the processing and release of statistical data must be managed carefully. Key statistics, such as unemployment rates, influ-

---

[5]For more on survey methodology and postsurvey editing, see, for example, Lars Lyberg et al. 1997. *Survey Measurement & Process Quality*. John Wiley & Sons, New York; and Brenda G. Cox et al. 1995. *Business Survey Methods*, John Wiley & Sons, New York. For more information on computer-assisted survey information collection (CASIC), see Mick P. Couper et al. 1998. *Computer Assisted Survey Information Collection*. John Wiley & Sons, New York.

**BOX 1.3
Survey Interview Methods**

- *Computer-Assisted Personal Interviewing (CAPI).* In CAPI, computer software guides the interviewer through a set of questions. Subsequent questions may depend on answers to previous questions (e.g., a respondent will be asked further questions about children in the household only if he/she indicates the presence of children). Questions asked may also depend on the answers given in prior interviews (e.g., a person who reports being retired will not be repeatedly asked about employment at the outset of each interview except to verify that he or she has not resumed employment). Such questions, and the resulting data captured, may also be hierarchical in nature. In a household survey, the responses from each member of the household would be contained within a household file. The combination of all of these possibilities can result in a very large number of possible paths through a survey instrument. CAPI software also may contain features to support case management.
- *Computer-Assisted Telephone Interviewing (CATI).* CATI is similar in concept to CAPI but supports an interviewer working by telephone rather than interviewing in person. CATI software may also contain features to support telephone-specific case management tasks, such as call scheduling.[1]
- *Computer-Assisted Self-Interviewing (CASI).* The person being interviewed interacts directly with a computer device. This technique is used when the direct involvement of a person conducting the interview might affect answers to sensitive questions. For instance, audio CASI, where the respondent responds to spoken questions, is used to gather mental health data in the NHANES.[2] The technique can also be useful for gathering information on sexual activities and illicit drug use.
- *Paper-and-Pencil Interviewing (PAPI).* Paper questionnaires, which pre-date computer-aided techniques, continue to be used in some surveys. Such questionnaires are obviously more limited in their ability to adapt or select questions based on earlier responses than the methods above, and they entail additional work (keying in responses prior to analysis). It may still be an appropriate method in certain cases, particularly where surveys are less complex, and it continues to be relied on as surveys shift to computer-aided methods. PAPI questionnaires have a smaller number of paths than computer-aided questionnaires; design and testing are largely a matter of formulating the questions themselves.

---

[1] The terms "CATI" and "CAPI" have specific, slightly different meanings when used by the Census Bureau. Field interviewers using a telephone from their home and a laptop are usually referred to as using CAPI, and only those using centralized telephone facilities are said to use CATI.

[2] The CASI technique is a subset of what is frequently referred to as computerized self-administered questionnaires, a broader category that includes data collection using Touch-Tone phones, mail-out-and-return diskettes, or Web forms completed by the interviewee.

ence business decisions and the financial markets, so it is critical that the correct information be released at the designated time and not earlier or later. Tight controls over the processes associated with data release are required. These stringent requirements also necessitate such measures as protection against attack of the database servers used to generate the statistical reports and the Web servers used to disseminate the final results. Process integrity and information system security research questions are discussed in the Chapter 2 section "Trustworthiness of Information Systems."

Creation and Dissemination of Statistical Products

Data are commonly released in different forms: as key statistics (e.g., the unemployment rate), as more extensive tables that summarize the survey data, and as detailed data sets that users can analyze themselves. Historically, most publicly disseminated data were made available in the form of printed tables, whereas today they are increasingly available in a variety of forms, frequently on the Internet. Tables from a number of surveys are made available on Web sites, and tools are sometimes provided for making queries and displaying results in tabular or graphical form. In other cases, data are less accessible to the nonexpert user. For instance, some data sets are made available as databases or flat-text files (either downloadable or on CD-ROM) that require additional software and/or user-written code to make use of the data.

A theme throughout the workshop was how to leverage IT to provide appropriate and useful access to a wide range of customers. A key consideration in disseminating statistical data, especially to the general public, is finding ways of improving its usability—creating a system that allows people, whether high school students, journalists, or market analysts, to access the wealth of statistical information that the government creates in a way that is useful to them. The first difficulty is simply finding appropriate data—determining which survey contains data of interest and which agencies have collected this information. An eventual goal is for users not to need to know which of the statistical agencies produced what data in order to find them; this and other data integration questions are discussed in the Chapter 2 section "Metadata." Better tools would permit people to run their own analyses and tabulations online, including analyses that draw on data from multiple surveys, possibly from different agencies.

Once an appropriate data set has been located, a host of other issues arise. There are challenges for both technological and statistical literacy in using and interpreting a data set. Several usability considerations are discussed in the Chapter 2 section "Human-Computer Interaction." Users also need ways of accessing and understanding what underlies the statistics, including the definitions used (a metadata issue, discussed in the Chapter 2 section "Metadata"). More sophisticated users will want to be able to create their own tabulations. For example, household income information might be available in pretabulated form by zip code, but a user might want to examine it by school district.

Because they contain information collected from individuals or organizations under a promise of confidentiality, the raw data collected from surveys are not publicly released as is or in their entirety; what is released is generally limited in type or granularity. Because this information is made available to all, careful attention must be paid to processing the data sets to reduce the chance that they can be used to infer information about individuals. This requirement is discussed in some detail in the Chapter 2 section "Limiting Disclosure." Concerns include the loss of privacy as a result of the release of confidential information as well as concerns about the potential for using confidential information to take administrative or legal action.[6]

However, microdata sets, which contain detailed records on individuals, may be made available for research use under tightly controlled conditions. The answers to many research questions depend on access to statistical data at a level finer than that available in publicly released data sets. How can such data be made available without compromising the confidentiality of the respondents who supplied the data? There are several approaches to address this challenge. In one approach, before they are released to researchers, data sets can be created in ways that de-identify records yet still permit analyses to be carried out. Another approach is to bring researchers in as temporary statistical agency staff, allowing them to access the data under the same tight restrictions that apply to other federal statistical agency employees. The section "Limiting Disclosure" in Chapter 2 takes up this issue in more detail.

ORGANIZATION OF THE FEDERAL STATISTICAL SYSTEM

The decentralized nature of the federal statistical system, with its more than 70 constituent agencies, has implications for both the efficiency of statistical activities and the ease with which users can locate and use

---

[6]The issue of balancing the needs for confidentiality of individual respondents with the benefits of accessibility to statistical data has been explored at great length by researchers and the federal statistical agencies. For a comprehensive examination of these issues see National Research Council and Social Science Research Council. 1993. *Private Lives and Public Policies*, George T. Duncan, Thomas B. Jabine, and Virginia A. deWolf, eds. National Academy Press, Washington, D.C.

INTRODUCTION AND CONTEXT 11

federal statistical data. Most of the work of these agencies goes on without any specific management attention by the Office of Management and Budget (OMB), which is the central coordinating office for the federal statistical system. OMB's coordinating authority spans a number of areas and provides a number of vehicles for coordination. The highest level of coordination is provided by the Interagency Council on Statistical Policy. Beyond that, a number of committees, task forces, and working groups address common concerns and develop standards to help integrate programs across the system. The coordination activities of OMB focus on ensuring that priority activities are reflected in the budgets of the respective agencies; approving all requests to collect information from 10 or more respondents (individuals, households, states, local governments, business);[7] and setting standards to ensure that agencies use a common set of definitions, especially in key areas such as industry and occupational classifications, the definition of U.S. metropolitan areas, and the collection of data on race and ethnicity.

In addition to these high-level coordination activities, strong collaborative ties—among agencies within the government as well as with outside organizations—underlie the collection of many official statistics. Several agencies, including the Census Bureau, the Bureau of Labor Statistics, and the National Agriculture Statistical Service, have large field forces to collect data. Sometimes, other agencies leverage their field-based resources by contracting to use these resources; state and local governments also perform statistical services under contracts with the federal government. Agencies also contract with private organizations such as Research Triangle Institute (RTI), Westat, National Opinion Research Center (NORC), and Abt Associates, to collect data or carry out surveys. (When surveys are contracted out, the federal agencies retain ultimate responsibility for the release of data from the surveys they conduct, and their contractors operate under safeguards to protect the confidentiality of the data collected.)

Provisions protecting confidentiality are also decentralized; federal statistical agencies must meet the requirements specified in their own particular legislative provisions. While some argue that this decentralized approach leads to inefficiencies, past efforts to centralize the system have run up against concerns that establishing a single, centralized statistical office could magnify the threat to privacy and confidentiality. Viewing the existence of multiple sets of rules governing confidentiality as a

---

[7]This approval process, mandated by the Paperwork Reduction Act of 1995 (44 U.S.C. 3504), applies to government-wide information-collection activities, not just statistical surveys.

barrier to effective collaboration and data sharing for statistical purposes, the Clinton Administration has been seeking legislation that, while maintaining the existing distributed system, would establish uniform confidentiality protections and permit limited data sharing among certain designated "statistical data center" agencies.[8] As a first step toward achieving this goal, OMB issued the Federal Statistical Confidentiality Order in 1997. The order is aimed at clarifying and harmonizing policy on protecting the confidentiality of persons supplying statistical information, assuring them that the information will be held in confidence and will not be used against them in any government action.[9]

In an effort to gain the benefits of coordinated activities while maintaining the existing decentralized structures, former OMB Director Franklin D. Raines posed a challenge to the Interagency Council on Statistical Policy (ICSP) in 1996, calling on it to implement what he termed a "virtual statistical agency." In response to this call, the ICSP identified three broad areas in which to focus collaborative endeavors:

- *Programs.* A variety of programs and products have interagency implications—an example is the gross domestic product, a figure that the Bureau of Economic Analysis issues but that is based on data from agencies in different executive departments. Areas for collaboration on statistical programs include establishing standards for the measurement of income and poverty and addressing the impacts of welfare and health care reforms on statistical programs.
- *Methodology.* The statistical agencies have had a rich history of collaboration on methodology; the Federal Committee on Statistical Methodology has regularly issued consensus documents on methodological issues.[10] The ICSP identified the following as priorities for collaboration: measurement issues, questionnaire design, survey technology, and analytical issues.
- *Technology.* The ICSP emphasized the need for collaboration in the area of technology. One objective stood out from the others because it was of interest to all of the agencies: to make the statistical system more

---

[8]Executive Office of the President, Office of Management and Budget (OMB). 1998. *Statistical Programs of the United States Government.* OMB, Washington, D.C., p. 40.

[9]Office of Management and Budget, Office of Information and Regulatory Affairs. 1997. "Order Providing for the Confidentiality of Statistical Information," *Federal Register* 62(124, June 27):33043. Available online at <http://www.access.gpo.gov/index.html>.

[10]More information on the Federal Committee on Statistical Methodology and on access to documents covering a range of methodological issues is available online from <http://fcsm.fedstats.gov/>.

INTRODUCTION AND CONTEXT   13

consistent and understandable for nonexpert users, so that citizens would not have to understand how the statistical system is organized in order to find the data they are looking for. The FedStats Web site,[11] sponsored by the Federal Interagency Council on Statistical Policy, is an initiative that is intended to respond to this challenge by providing a single point of access for federal statistics. It allows users to access data sets not only by agency and program but also by subject.

A greater emphasis on focusing federal statistics activities and fostering increased collaboration among the statistical agencies is evident in the development of the President's FY98 budget. The budgeting process for the executive branch agencies is generally carried out in a hierarchical fashion—the National Center for Education Statistics, for example, submits its budget to the Department of Education, and the Department of Education submits a version of that to the Office of Management and Budget. Alternatively, it can be developed through a cross-cut, where OMB looks at programs not only within the context of their respective departments but also across the government to see how specific activities fit together regardless of their home locations. For the first time in two decades, the OMB director called for a statistical agency cross-cut as an integral part of the budget formulation process for FY98.[12] In addition to the OMB cross-cut, the OMB director called for highlighting statistical activities in the Administration's budget documents and, thus, in the presentation of the budgets to the Congress.

Underlying the presentations and discussions at the workshop was a desire to tap IT innovations in order to realize a vision for the federal statistical agencies. A prominent theme in the discussions was how to address the decentralized nature of the U.S. national statistical system through virtual mechanisms. The look-up facilities provided by the FedStats Web site are a first step toward addressing this challenge. Other related challenges cited by workshop participants include finding ways for users to conduct queries across data sets from multiple surveys, including queries across data developed by more than one agency—a hard problem given that each survey has its own set of objectives and definitions associated with the information it provides. The notion of a virtual statistical agency also applies to the day-to-day work of the agencies. Although some legislative and policy barriers, discussed above in relation

---

[11] Available online from <http://www.fedstats.gov>.
[12] Note, however, that it was customary to have a statistical-agency cross-cut in each budget year prior to 1980.

to OMB's legislative proposal for data sharing, limit the extent to which federal agencies can share statistical data, there is interest in having more collaboration between statistical agencies on their surveys.

INFORMATION TECHNOLOGY INNOVATION IN FEDERAL STATISTICS

Federal statistical agencies have long recognized the pivotal role of IT in all phases of their activity. In fact, the Census Bureau was a significant driver of innovation in information technology for many years:

- Punch-card-based tabulation devices, invented by Herman Hollerith at the Census Bureau, were used to tabulate the results of the 1890 decennial census;
- The first Univac (Remington-Rand) computer, Univac I, was delivered in 1951 to the Census Bureau to help tabulate the results of the 1950 decennial census;[13]
- The Film Optical Scanning Device for Input to Computers (FOSDIC) enabled 1960 census questionnaires to be transferred to microfilm and scanned into computers for processing;
- The Census Bureau led in the development of computer-aided interviewing tools; and
- It developed the Topologically Integrated Geographic Encoding and Referencing (TIGER) digital database of geographic features, which covers the entire United States.

Reflecting a long history of IT use, the statistical agencies have a substantial base of legacy computer systems for carrying out surveys. The workshop case study on the IT infrastructure supporting the National Crime Victimization Survey illustrates the multiple cycles of modernization that have been undertaken by statistical agencies (Box 1.4).

Today, while they are no longer a primary driver of IT innovation, the statistical agencies continue to leverage IT in fulfilling their missions. Challenges include finding more effective and efficient means of collecting information, enhancing the data analysis process, increasing the availability of data while protecting confidentiality, and creating more usable, more accessible statistical products. The workshop explored, and this report describes, some of the mission activities where partnerships be-

---

[13]See, e.g., J.A.N. Lee. 1996. "looking.back: March in Computing History," *IEEE Computer* 29 (3). Available online from <http://computer.org/50/looking/r30006.htm>.

BOX 1.4
Modernization of the Information Technology Used for the National Crime Victimization Survey

Steven Phillips of the Census Bureau described some key elements in the development of the system used to conduct the National Crime Victimization Survey (NCVS) for the Bureau of Justice Statistics. He noted that the general trend over the years has been toward more direct communication with the sponsor agency, more direct communication with the subject matter analysts, quicker turnaround, and opportunities to modify the analysis system more rapidly. In the early days, the focus was on minimizing the use of central processing unit (CPU) cycles and storage space, both of which were costly and thus in short supply. Because the costs of both have continued to drop dramatically, the effort has shifted from optimizing the speed at which applications run to improving the end product.

At the data collection end, paper-and-pencil interviewing was originally used. In 1986, Mini-CATI, a system that ran on Digital Equipment Corporation minicomputers, was developed, and the benefits of online computer-assisted interviewing began to be explored. In 1989, the NCVS switched to a package called Micro-CATI, a quicker, more efficient, PC-based CATI system, and in 1999 it moved to a more capable CATI system that provides more powerful authoring tools and better capabilities for exporting the survey data and tabulations online to the sponsor. As of 1999, roughly 30 percent of the NCVS sample was using CATI interviewing.

Until 1985 a large Univac mainframe was used to process the survey data. It employed variable-length files; each household was structured into one record that could expand or contract. All the data in the tables were created by custom code, and the tables themselves were generated by a variety of custom packages. In 1986, processing shifted to a Fortran environment.

In 1989, SAS (a software product of the SAS Institute, Inc.) began to be used for the NCVS survey. At that time a new and more flexible nested and hierarchical data file format was adopted. Another big advantage of moving to this software system has been the ease with which tables can be created. Originally, all of the statistical tables were processed on a custom-written table generator. It produced a large numbers of tables, and the Bureau of Justice Statistics literally cut and pasted—with scissors and mucilage—to create the final tables for publications. A migration from mainframe-based Fortran software to a full SAS/Unix processing environment was undertaken in the 1990s; today, all processing is performed on a Unix workstation, and a set of SAS procedures is used to create the appropriate tables. All that remains to produce the final product is to process these tables, currently done using Lotus 1-2-3, into a format with appropriate fonts and other features for publication.

tween the IT research community and the statistics community might be fostered.

IT innovation has been taking place throughout government, motivated by a belief that effective deployment of new technology could vastly enhance citizens' access to government information and significantly streamline current government operations. The leveraging of information technology has been a particular focus of efforts to reinvent government. For example, Vice President Gore launched the National Performance Review, later renamed the National Partnership for Reinventing Government, with the intent of making government work better and cost less. The rapid growth of the Internet and the ease of use of the World Wide Web have offered an opportunity for extending electronic access to government resources, an opportunity that has been identified and exploited by the federal statistical agencies and others. Individual agency efforts have been complemented by cross-agency initiatives such as FedStats and Access America for Seniors.[14] While government agency Web pages have helped considerably in making information available, much more remains to be done to make it easy for citizens to locate and retrieve relevant, appropriate information.

Chapter 2 of this report looks at a number of research topics that emerged from the discussions at the workshop—topics that not only address the requirements of federal statistics but also are interesting research opportunities in their own right. The discussions resulted in another outcome as well: an increased recognition of the potential of interactions between government and the IT research community. Chapter 3 discusses some issues related to the nature and conduct of such interactions. The development of a comprehensive set of specific requirements or of a full, prioritized research agenda is, of course, beyond the scope of a single workshop, and this report does not presume to develop either. Nor does it aim to identify immediate solutions or ways of funding and deploying them. Rather, it examines opportunities for engaging the information technology research and federal statistics communities in research activities of mutual interest.

---

[14]Access America for Seniors, a government-operated Web portal that delivers electronic information and services for senior citizens, is available online at <http://www.seniors.gov>.

2

Research Opportunities

Research opportunities explored in the workshop's panel presentations and small-group discussions are described in this chapter, which illustrates the nature and range of IT research issues—including human-computer interaction, database systems, data mining, metadata, information integration, and information security—that arise in the context of the work being conducted by the federal statistical agencies. The chapter also touches on two other challenges pertinent to the work of the federal statistical agencies—survey instruments and the need to limit disclosure of confidential information. This discussion represents neither a comprehensive examination of information technology (IT) challenges nor a prioritization of research opportunities, and it does not attempt to focus on the more immediate challenges associated with implementation.

HUMAN-COMPUTER INTERACTION

One of the real challenges associated with federal statistical data is that the people who make use of it have a variety of goals. There are, first of all, hundreds or thousands of specialists within the statistical system who manipulate the data to produce the reports and indices that government agencies and business and industry depend on. Then there are the thousands, and potentially millions, of persons in the population at large who access the data. Some users access statistical resources daily, others only occasionally, and many others only indirectly, through third parties, but all depend in some fashion on these resources to support important

> **BOX 2.1**
> **Some Policy Issues Associated with Electronic Dissemination**
>
> In her presentation at the workshop, Patrice McDermott, from OMB Watch, observed that if information suddenly began to be disseminated by electronic means alone, some people would no longer be able to access it. Even basic telephone service, a precursor for low-cost Internet access, is not universal in the United States. It is not clear that schools and libraries can fill the gap: schools are not open, for the most part, to people who do not have children attending them, and finding resources to invest in Internet access remains a challenge for both schools and public libraries. McDermott added that research by OMB Watch indicates that people see a substantial difference between being directed to a book that contains Census data and being helped to access and navigate through online information. Another issue is the burden imposed by the shifting of costs: if information is available only in electronic form, users and intermediaries such as libraries end up bearing much of the cost of providing access to it, including, for example, the costs of telecommunications, Internet service, and printing.

decisions. Federal statistics resources support an increasingly diverse range of users (e.g., high school students, journalists, local community groups, business market analysts, and policy makers) and tasks. The pervasiveness of IT, exemplified by the general familiarity with the Web interface, is continually broadening the user base.

Workshop participants observed, however, that many are likely to remain without ready access to information online, raising a set of social and policy questions (Box 2.1). However, over time, a growing fraction of potential users can be expected to gain network access, making it increasingly beneficial to place information resources online, together with capabilities that support their interpretation and enhance the statistical literacy of users. In the meantime, online access is being complemented by published sources and by the journalists, community groups, and other intermediaries who summarize and interpret the data.

The responsibility of a data product designer or provider does not end with the initial creation of that product. There are some important human-computer interaction (HCI) design challenges in supporting a wide range of users. A key HCI design principle is "know thy user"; various approaches to learning about and understanding user abilities and needs are discussed below. Besides underscoring the need to focus on users, workshop participants pointed to some specific issues: universal access, support for users with limited statistical literacy, improved visualization techniques, and new modes of interacting with data. These are discussed in turn below.

User Focus

Iterative, user-centered design and testing are considered crucial to developing usable and useful information products. A better understanding of typical users and the most common tasks they perform, which could range from retrieving standard tables to building sophisticated queries, would facilitate the design of Web sites to meet those users' needs. One important approach discussed at the workshop is to involve the user from the start, through various routine participatory activities, in the design of sites. The capture of people's routine interactions with online systems to learn what users are doing, what they are trying to do, what questions they are asking, and what problems they are having allows improving the product design. If, for example, a substantial number of users are seen to ask the same question, the system should be modified to ensure that the answer to this question is easily available—an approach analogous to the "frequently asked questions" concept. Customer or market surveys can also be used in conjunction with ongoing log and site analyses to better understand the requirements of key user groups. There are many techniques that do not associate data with individuals and so are sensitive to privacy considerations.[1] For example, collecting frequent queries requires aggregation only at the level of the site, not of the individual. Where individual-level data are useful, they could be made anonymous.

Universal Access

The desire to provide access to statistical information for a broad range of citizens raises concerns about what measures must be taken to ensure universal access.[2] Access to computers, once the province of a small number of expert programmers, now extends to a wider set of computer-literate users and an even larger segment of the population sufficiently skilled to use the Web to access information. The expanding audience for federal statistical data represents both an opportunity and a challenge for information providers.

---

[1]Data on user behavior must be collected and analyzed in ways that are sensitive to privacy concerns and that avoid, in particular, tracking the actions of individuals over time (though this inhibits within-subject analyses). There are also the matters related to providing appropriate notice and obtaining consent for such monitoring.

[2]This term, similar to the more traditional label "universal service," also encompasses economic and social issues related to the affordability of access services and technology, as well as the provision of access through community-based facilities, but these are not the focus of this discussion.

Universality considerations apply as well to the interfaces people use to access information. The Web browser provides a common interface across a wide range of applications and extends access to a much larger segment of the population (anyone with a browser). However, the inertia associated with such large installed software bases tends to slow the implementation of new interface technologies. During the workshop, Gary Marchionini argued that adoption of the Web browser interface has locked in a limited range of interactions and in some sense has set interface design back several years. A key challenge in ensuring universal access is finding upgrade trajectories for interfaces that maximize access across the broadest possible audience.[3]

Providing access to all citizens also requires attention to the diverse physical needs of users. Making every Web site accessible to everyone requires more than delivering just a plain-text version of a document, because such a version lacks the richness of interaction offered by today's interfaces. Some work is already being done; vendors of operating systems, middleware, and applications provide software hooks that support alternative modes of access. The World Wide Web Consortium is establishing standards and defining such hooks to increase the accessibility of Web sites.

Another dimension of universal access is supporting users whose systems vary in terms of hardware performance, network connection speed, and software. The installed base of networked computers ranges from Intel 80286 processors using 14.4-kbps modems to high-performance computers with optical fiber links that are able to support real-time animation. That variability in the installed base presents a challenge in designing new interfaces that are also compatible with older systems and software.

Literacy, Visualization, and Perception

Given the relatively low level of numerical and statistical literacy in the population at large, it becomes especially important to provide users with interfaces that give them useful, meaningful information. Providing data with a bad interface that does not allow users to interpret data sensibly may be worse than not providing the data at all, because the bad interface frustrates nonexpert users and wastes their time. The goal is to provide not merely a data set but also tools that allow making sense of the data. Today, most statistical data is provided in tabular form—the form

---

[3]See Computer Science and Telecommunications Board, National Research Council. 1997. *More Than Screen Deep: Toward Every-Citizen Interfaces to the Nation's Information Infrastructure.* National Academy Press, Washington, D.C.

of presentation with which the statistical community has the longest experience. Unfortunately, although it is well understood by both statisticians and expert users, this form of presentation has significant limitations. Tables can be difficult for unsophisticated users to interpret, and they do not provide an engaging interface through which to explore statistical survey data. Also, the types of analyses that can be conducted using summary tables are much more limited than those that can be conducted when access to more detailed data is provided. Workshop participants pointed to the challenge of developing more accessible forms of presentation as central to expanding the audience for federal statistical data.

Statistics represent complex information that might be thought of as multimedia. Even data tables, when sufficiently large, do not lend themselves to display as simple text. Many of the known approaches to multimedia—such as content-based indexing and retrieval—may be applicable to statistical problems as well. Visualization techniques, such as user-controlled graphical displays and animations, enable the user to explore, discover, and explain trends, outliers, gaps, and jumps, allowing a better understanding of important economic or social phenomena and principles. Well-designed two-dimensional displays are effective for many tasks, but researchers are also exploring three-dimensional and immersive displays. Advanced techniques such as parallel coordinates and novel coding schemes, which complement work being done on three-dimensional and immersive environments, are also worthy of study.

Both representation (what needs to be shown to describe a given set of data) and control (how the user interacts with a system to determine what is displayed) pose challenges. Statisticians have been working on the problem of representation for a very long time. Indeed a statistic itself is a very concise condensation of a very large collection of information. More needs to be done in representing large data sets so that users who are not sophisticated in statistical matters can obtain, in a fairly compact way, the sense of the information in large collections of data. Related to this is the need to provide users with appropriate indications of the effects of sampling error.

Basic human perceptual and cognitive abilities affect the interpretation of statistical products. Amos Tversky and others have identified pervasive cognitive illusions, whereby people try to see patterns in random data.[4] In the workshop presentation by Diane Schiano, evidence

---

[4]See A. Tversky and D.M. Kahneman. 1974. "Judgement Under Uncertainty: Heuristics and Biases," *Science* 125:1124-1131. One such heuristic/bias is the perception of patterns in random scatter plots. See W.S. Cleveland and R. McGill. 1985. "Graphical Perception and Graphical Methods for Analyzing Scientific Data," *Science* 229 (August 30):828-833.

was offered of pervasive perceptual illusions that occur in even the simplest data displays. People make systematic errors in estimating the angle of a single line in a simple two-dimensional graph and in estimating the length of lines and histograms. These are basic perceptual responses that are not subject to cognitive overrides to correct the errors. As displays become more complex, the risk of perceptual errors grows accordingly. Because of this, three-dimensional graphics are often applied when they should not be, such as when the data are only two-dimensional. More generally, because complex presentations and views can suggest incorrect conclusions, simple, consistent displays are generally better.

The interpretation of complex data sets is aided by good exploratory tools that can provide both an overview of the data and facilities for navigating through them and zooming in (or "drilling down") on details. To illustrate the navigation challenge, Cathryn Dippo of the Bureau of Labor Statistics noted that the Current Population Survey's (CPS's) typical monthly file alone contains roughly 1,000 variables, and the March file contains an additional 3,000. Taking into account various supplements to the basic survey, the CPS has 20,000 to 25,000 variables, a number that rapidly becomes confusing for a user trying to interpret or even access the data. That figure is for just one survey; the surveys conducted by the Census Bureau contain some 100,000 variables in all.

Underscoring the importance of providing users with greater support for interaction with data, Schiano pointed to her research that found that direct manipulation through dynamic controls can help people correct some perceptual illusions associated with data presentation. Once users are allowed to interact with an information object and to choose different views, perception is vastly improved. Controls in common use today are limited largely to scrolling and paging through fairly static screens of information. However, richer modes of control are being explored, such as interfaces that let the user drag items around, zoom in on details, and aggregate and reorder data. The intent is to allow users to manipulate data displays directly in a much more interactive fashion.

Some of the most effective data presentation techniques emerging from human-computer interaction research involve tightly coupled interactions. For example, when the user moves a slider (a control that allows setting the value of a single variable visually), that action should have an immediate and direct effect on the display—users are not satisfied by an unresponsive system. Building systems that satisfy these requirements in the Web environment, where network communications latency delays data delivery and makes it hard to tightly couple a user action and the resulting display, is an interesting challenge. What, for example, are the optimal strategies for allocating data and processing between the client and the server in a networked environment in order to support this kind of interactivity?

Two key elements of interactivity are the physical interface and the overall style of interaction. The trend in physical interfaces has been toward a greater diversity of devices. For example, a mouse or other two-dimensional pointing device supplements keyboard input in desktop computing, while a range of three-dimensional interaction devices are used in more specialized applications. Indeed, various sensors are being developed that offer enhanced direct manipulation of data. One can anticipate that new ways of interacting will become commonplace in the future. How can these diverse and richer input and output devices be used to disseminate statistical information better? The benefits of building more flexible, interactive systems must be balanced against the risk that the increased complexity can lead unsophisticated users to draw the wrong conclusions (e.g., when they do not understand how the information has been transformed by their interactions with it).

Also at work today is a trend away from static displays toward what Gary Marchionini termed "hyperinteraction," which leads users to expect quick action and instant access to large quantities of information by pointing and clicking across the Web or by pressing the button on a TV remote control. An ever-greater fraction of the population has such expectations, affecting how one thinks about disseminating statistical information.

DATABASE SYSTEMS

Database systems cover a range of applications, from the large-scale relational database systems widely used commercially, to systems that provide sophisticated statistical tools and spreadsheet applications that provide simple data-manipulation functionality along with some analysis capability. Much of the work today in the database community is motivated by a commercial interest in combining transactions, analysis, and mining of multiple databases in a distributed environment. For example, data warehouse environments—terabyte or multiterabyte systems that integrate data from various locations—replicate transactions databases to support problem solving and decision making. Workshop participants observed that the problems of other user communities, such as the federal statistics community, can be addressed in this fashion as well.

Problems cited by the federal statistics community include legacy migration, information integration across heterogeneous databases, and mining data from multiple sources. These challenges, perhaps more mundane than the splashier Web development activities that many IT users are focused on, are nonetheless important. William Cody noted in the workshop that the database community has not focused much on these hard problems but is now increasingly addressing them in conjunction with its application partners. Commercial systems are beginning to address these needs.

Today's database systems do not build in all of the functionality to perform many types of analysis. There are several approaches to enhancing functionality, each with its advantages and disadvantages. Database systems can be expanded in an attempt to be all things to all people, or they can be constructed so that they can be extended using their own internal programming language. Another approach is to give users the ability to extract data sets for analysis using other tools and application languages. Researchers are exploring what functions are best incorporated in databases, looking at such factors as the performance trade-offs between the overhead of including a function inside a database and the delay incurred if a function must be performed outside the database system or in a separate database system.

Building increased functionality into database systems offers the potential for increasing overall processing efficiency, Cody observed. There are delays inherent in transferring data from one database to another; if database systems have enhanced functionality, processing can be done on a real-time or near-real-time basis, allowing much faster access to the information. Built-in functionality also permits databases to perform integrated tasks on data inside the database system. Also, relational databases lend themselves to parallelization, whereas tools external to databases have not been built to take as much advantage of it. Operations that can be included in the database engine are thus amenable to parallelization, allowing parallel processing computing capabilities to be exploited.

Cody described the likely evolution over the coming years of an interactive, analytic data engine, which has as its core a database system enriched with new functions. Users would be able to interact with the data more directly through visualization tools, allowing interactive data exploration. This concept is simple, but selecting and building the required set of basic statistical operations into database systems and creating the integration tools needed to use a workstation to explore databases interactively are significant challenges that will take time. Statistics-related operations that could be built into database systems include the following:

- *Data-mining operations.* By bringing data-mining primitives into the database, mining operations can occur automatically as data are collected in operational systems and transferred into warehousing systems rather than waiting until later, after special data sets have been constructed for data mining.
- *Enhanced statistical analysis.* Today, general-purpose relational database systems (as opposed to database systems specifically designed

for statistical analysis) for the most part support only fairly simple statistical operations. A considerable amount of effort is being devoted to figuring out which additional statistical operators should and could be included in evolving database systems. For example, could one perform a regression or compute statistical measures such as covariances and correlations directly in the database?

- *Time series operators.* The ability to conduct a time-series analysis within a database system would, for example, allow one to derive a forecast based on the information coming in real time to a database.
- *Sampling.* Sampling design is a sophisticated practice. Research is addressing ways to introduce sampling into database systems so that the user can make queries based on samples and obtain confidence limits around these results. While today's database systems use sampling during the query optimization process to estimate the result sizes of intermediate tables, sampling operators are not available to the end-user application. SQL, which is the standard language used to interact with database systems, provides a limited set of operations for aggregating data, although this has been augmented with the recent addition of new functionality for online analytical processing.

Additional support for statistical operations and sampling would allow, for example, estimating the average value of a variable in a data set containing millions of records by requesting that the database itself take a sample and calculate its average. The direct result, without any additional software to process the data, would be the estimated mean together with some confidence limit that would depend on the variance and the sample size.

Before the advent of object-relational database systems, which add object-oriented capabilities to relational databases, adding such extensions would generally have required extensive effort by the database vendor. Today, object-relational systems make it easier for third parties, as well as sophisticated users, to add both new data types and new operations into a database system. Since it is probably not reasonable to push all of the functionality of a statistical analysis product such as SAS into a general-purpose database system, a key challenge is to identify particular aggregation and sampling techniques and statistical operations that would provide the most leverage in terms of increasing both performance and functionality.

DATA MINING

Data mining enables the use of historical data to support evidence-based decision making—often without the benefit of explicitly stated statistical hypotheses—to create algorithms that can make associations that were not obvious to the database user. Ideas for data mining have been explored in a wide variety of contexts. In one example, researchers at Carnegie Mellon University studied a medical database containing several hundred medical features of some 10,000 pregnant women over time. They applied data-mining techniques to this collection of historical data to derive rules that better predict the risk of emergency caesarian sections for future patients. One pattern identified in the data predicts that when three conditions are met—no previous vaginal delivery, an abnormal second-trimester ultrasound reading, and the infant malpresenting—the patient's risk of an emergency caesarian section rises from a base rate of about 7 percent to approximately 60 percent.[5]

Data mining finds use in a number of commercial applications. A database containing information on software purchasers (such as age, income, what kind of hardware they own, and what kinds of software they have purchased so far) might be used to forecast who would be likely to purchase a particular software application in the future. Banks or credit card companies analyze historical data to identify customers that are likely to close their accounts and move to another service provider; predictive rules allow them to take preemptive action to retain accounts. In manufacturing, data collected over time from manufacturing processes (e.g., records containing various readings as items move down a production line) can be used by decision makers interested in process improvements in a production facility.

Both statisticians and computer scientists make use of some of the same data-mining tools and algorithms; researchers in the two fields have similar goals but somewhat different approaches to the problem. Statisticians, much as they would before beginning any statistical analysis, seek through interactions with the data owner to gain an understanding of how and why the data were collected, in part to make use of this information in the data mining and in part to better understand the limitations on what can be determined by data mining. The computer scientist, on the other hand, is more apt to focus on discovering ways to efficiently manipulate large databases in order to rapidly derive interesting or indicative trends and associations. Establishing the statistical validity of these methods and discoveries may be viewed as something that can be done at a later stage. Sometimes information on the conditions and circumstances under which the data were collected may be vague or even nonexistent, making it difficult to provide strong statistical justification for choosing

---

[5]This example is described in more detail in Tom M. Mitchell. 1999. "Machine Learning and Data Mining," *Communications of the ACM* 47(11).

particular data-mining tools or to establish the statistical validity of patterns identified from the mining; the statistician is arguably better equipped to understand the limitations of employing data mining in such circumstances. Statisticians seek to separate structure from noise in the data and to justify the separation based on principles of statistical inference. Similarly, statisticians approach issues like subsampling methodology as a statistical problem.

Research on data mining has been stimulated by the growth in both the quantity of data that is being collected and in the computing power available for analyzing it. At present, a useful set of first-generation algorithms has been developed for doing exploratory data analysis, including logistic regression, clustering, decision-tree methods, and artificial-neural-net methods. These algorithms have already been used to create a number of applications; at least 50 companies today market commercial versions of such analysis tools.

One key research issue is the scalability of data-mining algorithms. Mining today frequently relies on approaches such as selecting subsets of the data (e.g., by random sampling) and summarizing them, or deriving smaller data sets by methods other than selecting subsets (e.g., to perform a regression relating two variables, one might divide the data into 1,000 subgroups and perform the regression on each group, yielding a derived subset consisting of 1,000 sets of regression coefficients). For example, to mine a 4-terabyte database, one might do the following: sample it down to 200 gigabytes, aggregate it to 80 gigabytes, and then filter the result down to 10 gigabytes.

A relatively new area for data mining is multimedia data, including maps, images, and video. These are much more complex than the numerical data that have traditionally been mined, but they are also potentially rich new sources of information. While existing algorithms can sometimes be scaled up to handle these new types of data, mining them frequently requires completely new methods. Methods to mine multimedia data together with more traditional data sources could allow one to learn something that had not been known before. To use the earlier example, which involved determining risk factors in pregnancy, one would analyze not only the traditional features such as age (a numerical field) and childbearing status (a Boolean field) but also more complex multimedia features such as videosonograms and unstructured text notes entered by physicians. Another multimedia data-mining opportunity suggested at the workshop was to explore X-ray images (see Box 2.2) and numerical and text clinical data collected by the NHANES survey.

Active experimentation is an interesting research area related to data mining. Most analysis methods today analyze precollected samples of data. With the Internet and connectivity allowing researchers to easily

> **BOX 2.2**
> **National Health and Nutrition Examination Survey X-ray Image Archive**
>
> Lewis Berman of the National Center for Health Statistics presented some possible uses of the NHANES X-ray image archive. He described NHANES as the only nationally representative sampling of X rays and indicated that some effort had been made to make this set of data more widely available. For example, more than 17,000 X-ray cervical and lumbar spine images from NHANES II have been digitized.[1] In collaboration with the National Library of Medicine, these data are being made accessible online under controlled circumstances via Web tools, along with collateral data such as reported back pain at the time of the X ray. Other data sets that could also be useful to researchers include hand and knee films from NHANES III, a collection of hip X rays, and a 30-year compilation of electrocardiograms. NHANES data could also provide a resource that would allow the information technology and medical communities to explore issues ranging from multimedia data mining to the impact of image compression on the accuracy of automated diagnosis.
>
> ---
>
> [1]The images from NHANES II were scanned at 175 microns on a Lumisys Scanner. The cervical and lumbar spine images have a resolution of 1,463 × 1,755 × 12 bits (5 MB per image) and 2,048 × 2,487 × 12 bits (10 MB per image), respectively. Although the images are stored as 2 bytes/pixel, they capture only 12 bits of gray scale.

tap multiple databases, there is an opportunity to explore algorithms that would, after a first-pass analysis of an initial data set, search data sources on the Internet to collect additional data that might inform, test, or improve conjectures that are formed from the initial data set. In his presentation at the workshop, Tom Mitchell explored some of these implications of the Internet for data collection and analysis. An obvious opportunity is to make interview forms available on the Web and collect information from user-administered surveys. A more technically challenging opportunity is to make use of Web information that is already available. How might one use that very large, heterogeneous collection of data to augment the more carefully collected but smaller data sets that come from statistical surveys? For example, many companies in the United States have Web sites that provide information on current and new products, the company's location, and other information such as recruiting announcements. Mitchell cited work by his research group at Carnegie Mellon on extracting data from corporate Web sites to collect such information as where they are headquartered, where they have facilities, and what their economic sector is. Similarly, most universities have Web sites that describe their academic departments, degree programs, research activities, and faculty. Mitchell described a system that extracts information from the home pages of university faculty. It attempts to locate and identify faculty member Web sites by browsing university Web sites, and it extracts particular information on faculty members, such as their home department, the courses they teach, and the students they advise.[6]

METADATA

The term "metadata" is generally used to indicate the descriptions and definitions that underlie data elements. Metadata provides data about data. For example, what, precisely, is meant by "household" or "income" or "employed"? In addition to metadata describing individual data elements, there is a host of other information associated with a survey, also considered metadata, that may be required to understand and interpret a data set. These include memos documenting the survey, the algorithms[7] used to derive results from survey responses (e.g., how it is determined whether someone is employed), information on how surveys are constructed, information on data quality, and documentation of how the interviews are actually conducted (not just the questions asked but also the content of training materials and definitions used by interviewers in gathering the data). Workshop participants observed that better metadata and metadata tools and systems could have a significant impact on the usability of federal statistics, and they cited several key areas, discussed below.

Metadata, ranging from definitions of data fields to all other documentation associated with the design and conduct of a statistical survey, can be extensive. Martin Appel of the Census Bureau observed that attempts to manually add metadata have not been able to keep up with

---

[6]M. Craven, D. DiPasquo, D. Freitag, A. McCallum, T. Mitchell, K. Nigam, and S. Slattery. 1998. "Learning to Extract Symbolic Knowledge from the World Wide Web," *Proceedings of the 1998 National Conference on Artificial Intelligence* (July). Available online at <http://www.cs.cmu.edu/~tom/publications.html>.

[7]Simply including computer code as metadata may not be the most satisfactory method; even high-level language programs may not be useful as metadata. Another approach would be to use specification languages, which make careful statements about what computer code should do. These are more compact and more readable than typical computer code, although some familiarity with the specification language and comfort with its more formal nature are required. As with computer code itself, a description in a specification language cannot readily be interpreted by a nonexpert user, but it can be interpreted by a tool that can present salient details to nonexpert users. These languages are applicable not only to representing a particular computer program but also to representing larger systems, such as an entire statistical collection and processing system.

the volume of data that are generated. In particular, statistical data made available for analysis are frequently derived from calculations performed on other data, making the task of tying a particular data element to the appropriate metadata more complex. Tools for automatically generating and maintaining metadata as data sets are created, augmented, manipulated, and transformed (also known as self-documenting) could help meet this demand.

Even if fully satisfactory standards and tools are developed for use in future surveys, there remain legacy issues because the results of statistical surveys conducted in past decades are still of interest. For instance, the NHANES databases contain 30 years of data, during which time span similar but not identical questions were asked and evaluated, complicating the study of long-term health trends. Much work remains to provide a metadata system for these survey data that will permit their integration.

Another, related challenge is how to build tools that support the search and retrieval of metadata. A new user seeking to make sense of a Census data set may well need to know the difference between a "household" and a "family" or a "block group" and a "block" in order to make sense of that set. More generally, metadata are critical to help users make sense of data—for instance, what a particular piece of data means, how it was collected, and how much trust can be placed in it. The development of automatic display techniques that allow metadata associated with a particular data set to be quickly and easily accessed was identified as one area of need. For example, when a user examines a particular data cell, the associated metadata might be automatically displayed. At a minimum, drill-down facilities, such as the inclusion of a Web link in an online statistical report pointing to the relevant metadata, could be provided. Such tools should describe not only the raw data but also what sort of transformations were performed on them. Finally, as the next section discusses, metadata can be particularly important when one wishes to conduct an analysis across data from multiple sources.

INFORMATION INTEGRATION

Given the number of different statistical surveys and agencies conducting surveys, "one-stop shopping" for federal statistical data would make statistical data more accessible. Doing so depends on capabilities that allow analyzing data from multiple sources. The goal would be to facilitate both locating the relevant information across multiple surveys and linking it to generate new results. Several possible approaches were discussed at the workshop.

Metadata standards, including both standardized formats for describing the data as well as sets of commonly agreed-on meanings, are one key to fully exploiting data sets from multiple sources. Without them, for instance, it is very difficult to ascertain which fields in one data set correspond to which fields in the other set and to what extent the fields are comparable. While the framework provided by the recently developed XML standard, including the associated data-type definitions (DTDs), offers some degree of promise, work is needed to ensure that effective DTDs for federal statistical data sets are defined. XML DTDs, because they specify only certain structural characteristics of data, are only part of the solution; approaches for defining the semantics of statistical data sets also need to be developed. Standards do not, moreover, provide a solution for legacy data sets.

Another approach to information integration is to leverage the existing metadata, such as the text labels that describe the rows and columns in a statistical table or descriptions of how the data have been collected and processed, that accompany the data sets. Finding ways of using these metadata to represent and relate the contents of tables and databases so that analyses can be performed is an interesting area for further research.

The database community is exploring how to use database systems to integrate information originating from different systems throughout an organization (data warehousing). Database system developers are building tools that provide an interactive, analytical front end that integrates access to information in databases along with tools for visualizing the data. Research is being done on such things as data transformations and data cleaning and on how to model different data sources in an integrated way.

SURVEY INSTRUMENTS

The way in which data are collected is critical: without high-quality data up front, later work will have little value. Improved tools for administering surveys, whether they use paper and pencil, are computer-assisted, or are interviewee (end-user) administered, would also help. Discussions at the workshop suggested that a new generation of tools for developing surveys would offer statistical agencies greater flexibility in developing sound, comprehensive surveys. The current generation of tools is hard to use and requires that significant amounts of customized code be designed, written, and debugged. The complexity of the surveys sponsored by the federal government exceeds that of most other surveys, so it is unlikely that software to support this complex process will ever become mainstream. Workshop participants suggested that the federal government should for this reason consider consolidating its efforts to develop (or have others develop) such software. Some particular needs are associated with survey tools:

- *Improved survey software tools.* It would be useful to easily modify surveys that have already been developed or deployed; such modification can be difficult when extensive custom coding is required to create a survey instrument. High-level language tools (so-called fourth-generation languages), like those developed by the database industry, which demonstrate that families of sophisticated applications can be developed without requiring programmers to write extensive amounts of customized computer code, may also ease the task of developing surveys.

- *Flexibility in navigation.* Better software tools would, for example, permit users to easily back up to earlier answers and to correct errors. Heather Contrino, discussing the American Travel Survey CATI system, observed that if a respondent provides information about several trips during the trip section of the survey and then recalls another trip during the household section, it would be useful if the interviewer could immediately go back to a point in the survey where the new information should be captured and then proceed with the survey. The new CATI system used for the 1995 American Travel Survey provides some flexibility, but more would improve survey work. The issue, from an IT research perspective, is developing system designs that ensure internal consistency of the survey data acquired from subjects while also promoting more flexible interactions, such as adapting to respondents' spontaneous reports.

- *Improved ease of use.* Being able to visualize the flow of the questionnaire would be especially helpful. In complex interviews, an interviewer can lose his or her place and become disoriented, especially when following rarely used paths. This difficulty could be ameliorated by showing, for example, the current location in the survey in relation to the overall flow of the interview. Built-in training capabilities would also enhance the utility of future tools. Ideally, they should be able to coach the interviewer on how to administer the survey.

- *Monitoring the survey process.* Today, survey managers monitor the survey process manually. Tools for automatically monitoring the survey could be designed and implemented so that, as survey results are uploaded by the survey takers, status tables could be automatically produced and heuristic and statistical techniques used to detect abnormal conditions. Automated data collection would improve the timeliness of data collection and enhance monitoring efforts. While the data analyst is generally interested only in the final output from a survey instrument, the survey designer also wants information on the paths taken through the survey, including, for example, any information that was entered and then later modified. This is similar to the analyses of "click trace" that track user paths through Web sites.

- *On-the-fly response checking.* It would be useful to build in checks to identify inappropriate data values or contradictory answers immediately, as an interview is being conducted, rather than having to wait for post-interview edits and possibly incurring the cost and delay of a follow-up interview to correct the data. Past attempts to build in such checks are reported to have made the interview instruments run excessively slowly, so the checks were removed.

- *Improved performance.* Another dimension to the challenges of conducting surveys is the hardware platform. Laptops are the current platform of choice for taking a survey. However, the current generation of machines is not physically robust in the field, is too difficult to use, and is too heavy for many applications (e.g., when an interviewer stands in a doorway, as happens when a household is being screened for possible inclusion in a survey). Predictable advances in computer hardware will address size and shape, weight, and battery life problems while advances in processing speed will enable on-the-fly checking, as noted above. Continued commercial innovation in portable computer devices, building on the present generation of personal digital assistants, which provide sophisticated programmability, appears likely to provide systems suitable for many of these applications. It is, of course, a separate matter whether procurement processes and budgets can assimilate use of such products quickly.

- *New modes of interaction with survey instruments.* Another set of issues relates to the limitations of keyboard entry. While a keyboard is suitable for a telephone interview or an interview conducted inside someone's house, it has some serious limitations in other circumstances, such as when an interviewer is conducting an initial screening interview at someone's doorstep or in a driveway. Advances in speech-to-text technology might offer advantages for certain types of interviews, as might handwriting recognition capability, which is being made available in a number of computing devices today. Limited-vocabulary (e.g., "yes", "no," and numerical digits), speaker-independent speech recognition systems have been used for some time in survey work.[8] The technology envisioned here would provide speaker-independent capability with a less restricted vocabulary. With this technology it would be possible to capture answers in a much less intrusive fashion, which could lead to improvements in overall survey accuracy. Speech-to-text would also help reduce human intermediation if it could allow interviewees to interact directly with the survey instrument. There are significant research questions regarding the implications of different techniques for administering

---

[8]The Bureau of Labor Statistics started using this technology for the Current Employment Survey in 1992. See Richard L. Clayton and Debbie L.S. Winter. 1992. "Speech Data Entry: Results of a Test of Voice Recognition for Survey Data Collection," *Journal of Official Statistics* 8:377-388.

survey questionnaires, with some results in the literature suggesting that choice of administration technique can affect survey results significantly.[9] More research on this question, as well as on the impact of human intermediation on data collection, would be valuable.

LIMITING DISCLOSURE

Maintaining the confidentiality of respondents in data collected under pledges of confidentiality is an intrinsic part of the mission of the federal statistical agencies. It is this promise of protection against disclosure of confidential information—protecting individual privacy or business trade secrets—that convinces many people and businesses to comply willingly and openly with requests for information about themselves, their activities, and their organizations. Hence, there are strong rules in place governing how agencies may (and may not) share data,[10] and data that divulge information about individual respondents are not released to the public. Disclosure limitation is a research area that spans both statistics and IT; researchers in both fields have worked on the issue in the past, and approaches and techniques from both fields have yielded insights. While nontechnical approaches play a role, IT tools are frequently employed to help ease the tension between society's demands for data and the agencies' ability to collect information and maintain its confidentiality.

Researchers rely on analysis of data sets from federal statistical surveys, which are viewed as providing the highest-quality data on a number of topics, to explore many economic and social phenomena. While some of their analysis can be conducted using public data sets, some of it depends on information that could be used to infer information about individual respondents, including microdata, which are the data sets containing records on individual respondents. Statistical agencies must strike a balance between the benefits obtained by releasing information for legitimate research and the potential for unintended disclosures that could result from releasing information. The problem is more complicated than simply whether or not to release microdata. Whenever an agency releases statistical information, it is inherently disclosing some information about

---

[9]See, e.g., Sara Kiesler and Lee Sproull. 1986. "Response Effects in the Electronic Survey," *Public Opinion Quarterly* 50:243-253 and Wendy L. Richman, Sara Kiesler, Suzanne Weisband, and Fritz Drasgow. 1999. "A Meta-analytic Study of Social Desirability Distortion in Computer-Administered Questionnaires, Traditional Questionnaires, and Interviews," *Journal of Applied Psychology* 84(5, October):754-775.

[10]These rules were clarified and stated consistently in Office of Management and Budget, Office of Information and Regulatory Affairs. 1997. "Order Providing for the Confidentiality of Statistical Information," *Federal Register* 62(124, June 27):33043. Available online at <http://www.access.gpo.gov/index.html>.

the source of the data from which the statistics are computed and potentially making it easier to infer information about individual respondents.

Contrary to what is sometimes assumed, protecting data confidentiality is not as simple as merely suppressing names and other obvious identifiers. In some cases, one can re-identify such data using record linkage techniques. Record linkage, simply put, is the process of using identifying information in a given record to identify other records containing information on the same individual or entity.[11] For example, a set of attributes such as geographical region, sex, age, race, and so forth may be sufficient to identify individuals uniquely. Moreover, because multiple sources of data may be drawn on to infer identity, understanding how much can be inferred from a particular set of data is difficult. A simple example provided by Latanya Sweeney in her presentation at the workshop illustrates how linking can be used to infer identity (Box 2.3).

Both technical and nontechnical approaches have a role in improving researcher access to statistical data. Agencies are exploring a variety of nontechnical solutions to complement their technical solutions. For example, the National Center for Education Statistics allows researchers access to restricted-use data under strict licensing terms, and the National Center for Health Statistics (NCHS) recently opened a research data center that makes data files from many of its surveys available, both on-site and via remote access, under controlled conditions. The Census Bureau has established satellite centers for secured access to research data in partnership with the National Bureau of Economic Research, Carnegie Mellon University, and the University of California (at Berkeley and at Los Angeles), and it intends to open additional centers.[12] Access to data requires specific contractual arrangements aimed at safeguarding confidentiality, and de-identified public-use microdata user files can be accessed through third parties. For example, data from the National Crime Victimization Survey are made available through the Interuniversity Consortium for Political and Social Research (ICPSR) at the University of Michigan. Members of the research community are, of course, interested in finding less restrictive ways of giving researchers access to confidential data that do not compromise the confidentiality of that data.

---

[11]For an overview and series of technical papers on record linkage, see Committee on Applied and Theoretical Statistics, National Research Council and Federal Committee on Statistical Methodology, Office of Management and Budget. 1999. *Record Linkage Techniques—1997: Proceedings of an International Workshop and Exposition.* National Academy Press, Washington, D.C.

[12]See U.S. Census Bureau, Office of the Chief Economist, 1999. *Research Data Centers.* U.S. Census Bureau, Washington, D.C., last revised September 28. Available online at <http://www.census.gov/cecon/www/rdc.html>.

> **BOX 2.3**
> **Using External Data to Re-identify Personal Data**
>
> Removing names and other unique identification information is not sufficient to prevent re-identifying the individuals associated with a particular data record. Latanya Sweeney illustrated this point in her presentation at the workshop using an example of how external data sources can be used to determine the identity of the individuals associated with medical records. Hospitals and insurers collect information on individual patients. Because such data are generally believed to be anonymous once names and other unique identifiers have been removed, copies of these data sets are provided to researchers and sold commercially. Sweeney described how she re-identified these seemingly anonymous records using information contained in voter registration records, which are readily purchased for many communities.
>
> Voter registration lists, which provide information on name, address, and so forth, are likely to have three fields in common with de-identified medical records—zip code, birth date, and sex. How unique a link can be established using this information? In one community where Sweeney attempted to re-identify personal data, there are 54,805 voters. The range of possible birth dates (year, month, day) is relatively small—about 36,500 dates over 100 years—and so potentially can be useful in identifying individuals. In the community she studies, there is a concentration of people in their 20s and 30s, and birth date alone uniquely identifies about 12 percent of the community's population. That is, given a person's birth date and knowledge that the person lived in that community, one could uniquely identify him or her. Birth date and gender were unique for 29 percent of the voters, birth date and zip code, for 69 percent, and birth date and full postal code, for 97 percent.

Academic work on IT approaches to disclosure limitation has so far been confined largely to techniques for limiting disclosure resulting from release of a given data set. However, as the example provided by Sweeney illustrates, disclosure limitation must also address the extent to which released information can be combined with other, previously released statistical information, including administrative data and commercial and other publicly available data sets, to make inferences. Researchers have recognized the importance of understanding the impact on confidentiality of these external data sources, but progress has been limited because the problem is so complex. The issue is becoming more important for at least two reasons. First, the quantity of personal information being collected automatically is increasing rapidly (Box 2.4) as the Web grows and database systems become more sophisticated. Second, the statistical agencies, to meet the research needs of their users, are being asked to release "anonymized" microdata to support additional data analyses. As a result, a balancing act must be performed between the benefits obtained from

> **BOX 2.4**
> **Growth in the Collection of Personal Data**
>
> At the workshop, Latanya Sweeney described a metric she had developed to provide a sense of how the amount of personal data is growing. Her measure—disk storage per person, calculated as the amount of storage in the form of hard disks sold per year divided by the adult world population—is based on the assumption that access to inexpensive computers with very large storage capacities is enabling the collection of an increasing amount of personal data. Based on this metric, the several thousand characters of information that could be printed on an 8 1/2 by 11 inch piece of paper would have documented some 2 months of a person's life in 1983. The estimate seems reasonable: at that time such information probably would have been limited to that contained in school or employment records, the telephone calls contained on telephone bills, utility bills, and the like. By 1996, that same piece of paper would document 1 hour of a person's life. The growth can be seen in the increased amount of information contained on a Massachusetts birth certificate; it once had 15 fields of information but today has more than 100. Similar growth is occurring in educational data records, grocery store purchase logs, and many other databases, observed Sweeney. Projections for the metric in 2000, with 20-gigabyte drives widely available, are that the information contained on a single page would document less than 4 minutes of a person's life—information that includes image data, Web and Internet usage data, biometric data (gathered for health care, authentication, and even Web-based clothing purchases), and so on.

data release and the potential for unwanted disclosure that comes from linking with other databases. What is the disclosure effect, at the margin, of the release of a particular set of data from a statistical agency?

The issue of disclosure control has also been addressed in the context of work on multilevel security in database systems, in which the security authorization level of a user affects the results of database queries.[13] A simple disclosure control mechanism such as classifying individual records is not sufficient because of the possible existence of an inference channel whereby information classified at a level higher than that for which a user is cleared can be inferred by that user based on information at lower levels (including external information) that is possessed by that

---

[13] See National Research Council and Social Science Research Council. 1993. *Private Lives and Public Policies: Confidentiality and Accessibility of Government Statistics*. National Academy Press, Washington, D.C., pp. 150-151; and D.E. Denning et al. 1988. "A Multilevel Relational Data Model," *Proceedings of the 1987 IEEE Symposium on Research Security and Privacy*. IEEE Computer Society, Los Alamitos, Calif.

user. Such channels are, in general, hard to detect because they may involve a complex chain of inferences and because of the ability of users to exploit external data.[14]

Various statistical disclosure-limiting techniques have been and are being developed to protect different types of data. The degree to which these techniques need to be unique to specific data types has not been resolved. The bulk of the research by statistics researchers on statistical disclosure limitation has focused on tabular data, and a number of disclosure-limiting techniques have been developed to protect the confidentiality of individual respondents (including people and businesses), including the following:

- *Cell suppression*—the blanking of table entries that would provide information that could be narrowed down to too small a set of individuals;
- *Swapping*—exchanging pieces of information among similar individuals in a data set; and
- *Top coding*—aggregating all individuals above a certain threshold into a single top category. This allows, for example, hiding information about an individual whose income was significantly greater than the incomes of the other individuals in a given set that would otherwise appear in a lone row of a table.

However, researchers who want access to the data are not yet satisfied with currently available tabular data-disclosure solutions. In particular, some of these approaches rely on distorting the data in ways that can make it less acceptable for certain uses. For example, swapping can alter records in a way that throws off certain kinds of research (e.g., it can limit researchers' ability to explore correlations between various attributes).

While disclosure issues for tabular data sets have received the most attention from researchers, many other types of data are also released, both publicly and to more limited groups such as researchers, giving rise to a host of questions about how to limit disclosure. Some attention has been given to microdata sets and the creation of public-use microdata

---

[14]See T.F. Lunt, T.D. Garvey, X. Qian, and M.E. Stickel. 1994. "Type Overlap Relations and the Inference Problem," *Proceedings of the 8th IFIP WG 11.3 Working Conference on Database Security*, August; T.F. Lunt, T.D. Garvey, X. Qian, and M.E. Stickel. 1994. "Issues in Data-Level Monitoring of Conjunctive Inference Channels," *Proceedings of the 8th IFIP WG 11.3 Working Conference on Database Security*, August; and T.F. Lunt, T.D. Garvey, X. Qian, and M.E. Stickel. 1994. "Detection and Elimination of Inference Channels in Multilevel Relational Database Systems," *Proceedings of the IEEE Symposium on Research in Security and Privacy*, May 1993. For an analysis of the conceptual models underlying multilevel security, see Computer Science and Telecommunications Board, National Research Council. 1999. *Trust in Cyberspace*. National Academy Press, Washington, D.C.

files. The proliferation of off-the-shelf software for data linking and data combining appears to have raised concerns about releasing microdata. None of the possible solutions to this problem coming from the research community (e.g., random sampling, masking, or synthetic data generation) seems mature enough today to be adopted as a data release technique.

Digital geospatial data, including image data, are becoming more widely available and are of increasing interest to the research community. Opportunities for and interest in linking data sets by spatial coordinates can be expected to grow correspondingly. In many surveys, especially natural resources or environmental surveys, the subject matter is inherently spatial. And spatial data are instrumental in research in many areas, including public health and economic development. The confidentiality of released data based on sample surveys is generally protected by minimizing the chance that a respondent can be uniquely identified using demographic variables and other characteristics. The situations where sampling or observational units (e.g., person, household, business, or land plot) are linked with a spatial coordinate (e.g., latitude and longitude) or another spatial attribute (e.g., Census block or hydrologic unit) have been less well explored. Precise spatial coordinates for sampling or observational units in surveys are today generally considered identifying information and are thus excluded from the information that can be released with a public data set. Identification can also be achieved through a combination of less precise spatial attributes (e.g., county, Census block, hydrologic unit, land use), and care must be taken to ensure that including variables of this sort in a public data set will not allow individual respondents to be uniquely identified.

Techniques to limit information disclosure associated with spatial data have received relatively little attention, and research is needed on approaches that strike an appropriate balance between two opposing forces: (1) the need to protect the confidentiality of sample and observational units when spatial coordinates or related attributes are integral to the survey and (2) the benefits of using spatial information to link with a broader suite of information resources. Such approaches might draw from techniques currently used to protect the confidentiality of alphanumeric human population survey data. For example, random noise might be added to make the spatial location fuzzier, or classes of spatial attributes might be combined to create a data set with lower resolution. It is possible that the costs and benefits of methods for protecting the confidentiality of spatial data will vary from those where only alphanumeric data are involved. In addition, alternative paradigms making use of new information technologies may be more appropriate for problems specific to spatial data. One might, for instance, employ a behind-the-scenes mechanism for accurately combining spatial information where the linkage, such as the merging of spatial data sets, occurs in a confidential "space" to produce a product such as a map or a data set with summaries that do not disclose locations. In some cases, this might include a mechanism that implements disclosure safeguards.

A third, more general, issue is how to address disclosure limitation when multimedia data such as medical images are considered. Approaches developed for numerical tabular or microdata do not readily apply to images, instrument readings, text, or combinations of them. For example, how does one ensure that information gleaned from medical images cannot be used to re-identify records? Given the considerable interest of both computer scientists and statisticians in applying data-mining techniques to extract patterns from multimedia data, collaboration with computer scientists on disclosure-limiting techniques for these data is likely to be fruitful.

Few efforts have been made to evaluate the success of data release strategies in practice. Suppose for example, that a certain database is proposed for release. Could one develop an analytical technique to help data managers evaluate the potential for unwanted disclosure caused by the proposed release? The analysis would evaluate the database itself, along with meta-information about other known, released databases, so as to identify characteristics of additional external information that could cause an unwanted disclosure. It could be used to evaluate not only the particular database proposed for release but also the impact of that release on potential future releases of other databases. Several possible approaches were identified by workshop participants. First, one can further develop systematic approaches for testing the degree to which a particular release would identify individuals. Given that it is quite difficult to know the full scope of information available to a would-be "attacker," it might also be useful to develop models of the information available to and the behavior of someone trying to overcome attempts to limit disclosure and to use these models to test the effectiveness of a particular disclosure limitation approach.

Another approach, albeit a less systematic one, is to explore red teaming to learn how a given data set could be exploited (including by combining it with other, previously disclosed or publicly available data sets). Red teaming in this context is like red teaming to test information system security (a team of talented individuals is invited to probe for weaknesses in a system[15]), and the technique could benefit from collaboration with IT researchers and practitioners.

---

[15]A recent CSTB report examining defense command-and-control systems underscored the importance of frequent red teaming to assess the security of critical systems. See Computer Science and Telecommunications Board, National Research Council. 1999. *Realizing the Potential of C4I: Fundamental Challenges*. National Academy Press, Washington, D.C.

TRUSTWORTHINESS OF INFORMATION SYSTEMS

The challenge of building trustworthy (secure, dependable, and reliable) systems has grown along with the increasing complexity of information systems and their connectedness, ubiquity, and pervasiveness. This is a burgeoning challenge to the federal statistical community as agencies move to greater use of networked systems for data collection, processing, and dissemination. Thus, even as solutions are developed, the goal being pursued often appears to recede.[16]

There have been substantial advances in some areas of security and particular problems have been solved. For example, if one wishes to protect information while it is in transit on a network, the technology to do this is generally considered to be available.[17] Hence experts tend to agree that a credit card transaction over the Internet can be conducted with confidence that credit card numbers cannot be exposed or tampered with while they are in transit. On the other hand, there remain many difficult areas: for example, unlike securing information in transit, the problem of securing the information on the end systems has, in recent years, not received the attention that it demands. Protecting against disclosure of confidential information and ensuring the integrity of the collection, analysis, and dissemination process are critical issues for federal statistical agencies.

For the research community that depends on federal statistics, a key security issue is how to facilitate access to microdata sets without compromising their confidentiality. As noted above, the principal approach being used today is for researchers to relocate themselves temporarily to agency offices or one of a small number of physically secured data centers, such as those set up by the Census Bureau and the NCHS. Unfortunately, the associated inconveniences, such as the need for frequent travel, are cited by researchers as a significant impediment to working with microdata. Another possible approach being explored is the use of various security techniques to permit off-site access to data. NCHS is one agency that has established remote data access services for researchers. This raises several issues. For example, what is the trade-off between

---

[16]The recent flap over the proposed Federal Intrusion Detection Network (FIDnet) indicates that implementing security measures is more complicated in a federal government context.

[17]For a variety of reasons, including legal and political issues associated with restrictions that have been placed on the export of strong cryptography from the United States, these technologies are not as widely deployed as some argue they should be. See, e.g., Computer Science and Telecommunications Board, National Research Council. 1996. *Cryptography's Role in Securing the Information Society*. National Academy Press, Washington, D.C. These restrictions have recently been relaxed.

permitting off-site users to replicate databases to their own computers in a secure fashion for local analysis and permitting users to have secured remote access to external analysis software running on computers located at a secured center. Both approaches require attention to authentication of users and both require safeguards, technological or procedural, to prevent disclosure as a result of the microdata analysis.[18]

Another significant challenge in the federal statistics area is maintaining the integrity of the process by which statistical data are collected, processed, and disseminated. Federal statistics carry a great deal of authority because of the reputation that the agencies have developed—a reputation that demands careful attention to information security. Discussing the challenges of maintaining the back-end systems that support the electronic dissemination of statistics products, Michael Levi of the Bureau of Labor Statistics cited several demands placed on statistics agencies: systems that possess automated failure detection and recovery capabilities; better configuration management including installation, testing, and reporting tools; and improved tools for intrusion prevention, detection, and analysis.

As described above, the federal statistical community is moving away from manual, paper-and-pencil modes of data collection to more automated modes. This trend started with the use of computer-assisted techniques (e.g., CAPI and CATI) to support interviewers and over time can be expected to move toward more automated modes of data gathering, including embedded sensors for automated collection of data (e.g., imagine if one day the American Travel Survey were to use Global Positioning System satellite receivers and data recorders instead of surveys). Increasing automation increases the need to maintain the traceability of data to its source as the data are transferred from place to place (e.g., uploaded from a remote site to a central processing center) and are processed into different forms during analysis (e.g., to ensure that the processed data in a table in fact reflect the original source data). In other words, there is a greater challenge in maintaining process integrity—a chain of evidence from source to dissemination.

There are related challenges associated with avoiding premature data release. In some instances, data have been inadvertently released before the intended point in time. For example, the Bureau of Labor Statistics prematurely released part of its October 1998 employment report.

---

[18]A similar set of technical requirements arise in supporting the geographically dispersed workers who conduct field interviews and report the data that have been collected. See, for example, Computer Science and Telecommunications Board, National Research Council. 1992. *Review of the Tax Systems Modernization of the Internal Revenue Service.* National Academy Press, Washington, D.C.

According to press reports citing a statement made by BLS Commissioner Katharine G. Abraham, this happened when information was moved to an internal computer by a BLS employee who did not know it would thereupon be transferred immediately to the agency's World Wide Web site and thus be made available to the public.[19] The processes for managing data apparently depended on manual procedures. What kind of automated process-support tools could be developed to make it much more difficult to release information prematurely?

In the security research literature, problems and solutions are abstracted into a set of technologies or building blocks. The test of these building blocks is how well researchers and technologists can apply them to understand and address the real needs of customers. While there are a number of unsolved research questions in information security, solutions can in many cases be obtained through the application of known security techniques. Of course the right solution depends on the context; security design is conducted on the basis of knowledge of vulnerabilities and threats and the level of risk that can be tolerated, and this information is specific to each individual application or system. Solving real problems also helps advance more fundamental understanding of security; the constraints of a particular problem environment can force rethinking of the structure of the world of building blocks.

---

[19]John M. Berry. 1998. "BLS Glitch Blamed on Staff Error; Premature Release of Job Data on Web Site Boosted Stocks," *Washington Post*, November 7, p. H03.

3

Interactions for Information Technology Innovation in Federal Statistical Work

The workshop discussed the information technology (IT) requirements of the federal statistical agencies and the research questions motivated by those needs. In addition to articulating a sizable list of research topics, workshop participants made a number of observations about the nature of the relationship and interactions between the two communities. These observations are offered to illustrate the sorts of issues that arise in considering how to foster collaboration and interaction between the federal statistical agencies and the IT research community aimed at innovation in the work of the agencies.[1]

One obstacle discussed in the course of the workshop is that despite interest in innovation, there are insufficient connections between those who operate and develop government information systems or who run agency programs and those who conduct IT research. In particular, federal agencies, like most procurers of IT systems, tend to rely on what is available from commercial technology vendors or system integrators (or, in some cases, what can be developed or built in-house). A program aimed at bridging this gap, the National Science Foundation's (NSF's) Digital Government program, was launched in June 1998 to support research aimed at stimulating IT innovation in government. The premise of

---

[1] These observations should not be viewed as necessarily being conclusions of the study committee that organized the workshop. The committee's conclusions will be presented in the study's final report, to be published later in 2000.

this program is that by promoting interactions between innovators in government and those performing computing and communications research, it may be possible both to accelerate innovation in pertinent technical areas and to hasten the adoption of those innovations into agency infrastructure.

Building connections that address the needs and interests of both communities entails the establishment of appropriate mechanisms for collaboration between the IT research community and government IT managers. In principle, the right mechanisms can help federal program and IT acquisition managers interact with the IT research community without exposing operational users to unacceptable levels of risk. Also, incorporating new research ideas and technology into the operations of government agencies frequently requires spanning a gulf between the culture and practices of commercial systems integration and the research community.

Also relevant to the issue of innovation and risk in the context of government in general, and the federal statistical system in particular, is the value attached to the integrity of the federal statistics community and the trustworthiness of the results (relevant principles are summarized in Box 1.1). These are attributes that the agencies value highly and wish to preserve and that have led to a strong tradition of careful management. Such considerations could constrain efforts that experiment with new technologies in these activities.

Experience suggests that despite these potential constraints and inhibitors, both research and application communities stand to benefit from interaction. Introduction of new IT can enable organizations to optimize the delivery of existing capabilities. The full benefits of IT innovation extend further, as such innovation can enable organizations to do things in new ways or attain entirely new capabilities. Advances in IT research represent opportunities not only for increased efficiency but also for changes in the way government works, including the delivery of new kinds of services and new ways of interacting with citizens. Collaboration with government agencies also represents a significant opportunity for IT researchers to test new ideas—government applications are real and have texture, richness, and veracity that are not available in laboratory studies. Frequently, these applications are also of a much larger scale than that found in most research work.

While the workshop focused primarily on long-term issues, another benefit was the shedding of light on some short-term problems. Indeed, it is natural for people in an operational setting to focus on problems that need to be solved in the next year rather than on long-term possibilities. This suggests that focus on and investment in long-term challenges may be difficult. But in some respects, the near-term focus may be appropriate, since some of the information technologies and IT practices of the federal statistical agencies lag behind best industry practices. In an example illustrating the short-term, mundane challenges that consume considerable time and resources, one workshop presenter described the challenges posed by the need to install new software on several hundred laptop computers. In later discussions, it was pointed out that this was a problem that had already been solved in the marketplace; there are well-known techniques for disk imaging that allow initialization of thousands of computers. Underscoring the potential value of such interactions, informal feedback following the workshop suggested that the exposure to some cutting-edge computer science thinking stimulated subsequent discussion among some statistical agencies about their need for further modernization.

One factor that may be exacerbating many of the short- and long-term IT-related challenges is the decentralized nature of the federal statistical agencies, which makes it harder to establish a critical mass of expertise, investment, and experimental infrastructure. Another difficulty arises from the specialized requirements of federal statistical agencies. The market is limited for software for authoring and administering survey interviews of the complexity found in federal statistical surveys, which are quite expensive and are conducted only by government and a few other players. Workshop participants discussed how the federal government might consolidate its research and development efforts for this class of software. Several IT applications in this category were cited, including survey software, easy-to-use interfaces for displaying complex data sets, and techniques for limiting the disclosure of confidential information in databases.

Collaborative research, even within a discipline, is not always easy, and interdisciplinary work is harder still. Researchers at the workshop argued that in order for such collaboration to take place, both IT and statistics researchers would need to explore ways of tapping existing research programs or establishing new funding mechanisms.[2] Computer scientists do not typically think of going to one of the statistical agencies, and statisticians do not typically think about teaming with a computer scientist for their fellowship research. Both computer scientists and statisticians will find it easier to obtain funding for work in more traditional

---

[2]Workshop participants pointed to two NSF programs that could facilitate such collaborations if they were explicitly targeted toward such interactions. One is a fellows program in the Methodology, Measurement, and Statistics program that sends statisticians to various federal statistical agencies. The second is a Computer and Information Science and Engineering (CISE) directorate program that provides support for computer scientists to take temporary positions in federal agencies.

research directions. So, given all the additional difficulties associated with interdisciplinary work, particularly in academia, it is unlikely to occur without funding directed at collaborative work.[3] This, of course, was part of the impetus for the NSF Digital Government program.

More generally, a number of workshop participants acknowledged that involvement in application areas related to federal statistics offers significant opportunities for IT researchers. Each of the areas described in Chapter 2 was identified by participants as one where considerable benefits would be obtained from direct collaboration between IT and statistics researchers. A leading example is the area of information security. While some segments of the computer science community may be ambivalent about doing application-focused research, it is difficult to make real progress in information security without a specific application focus. A similarly large challenge is building easy-to-use systems that enable nonexpert users, who have diverse needs and capabilities, to access, view, and analyze data. Both the magnitude of the challenge itself and the opportunity to conduct research on systems used by a large pool of diverse users make these systems an attractive focus for research. Another particularly interesting issue discussed by workshop participants was the development of techniques to protect the confidentiality of spatial data.

---

[3]Participants in a workshop convened by CSTB that explored ways to foster interdisciplinary research on the economic and social impacts of information technology made similar observations. See Computer Science and Telecommunications Board, National Research Council. 1998. *Fostering Research on the Economic and Social Impacts of Information Technology: Report of a Workshop.* National Academy Press, Washington, D.C.

Appendix

Workshop Agenda and Participants

AGENDA

Tuesday, February 9, 1999

7:30 a.m.	Registration and Continental Breakfast
8:30	**Welcome** *William Scherlis*
8:45	**Keynote Address** *Thomas Kalil*, National Economic Council
9:15	**Panel 1: Case Studies** • National Health and Nutrition Examination Surveys, *Lewis Berman* • American Travel Study, *Heather Contrino* • Current Population Survey, *Cathryn Dippo* • National Crime Victimization Survey, *Denise Lewis* *Sallie Keller-McNulty*, Moderator
11:00	**Panel 2: Information Technology Trends and Opportunities** *Gary Marchionini, Tom Mitchell, Ravi S. Sandhu, William Cody, Clifford Neuman* (moderator)
12:30 p.m.	Lunch
1:30	**Panel 3: Study Design, Data Collection, and Data Processing** *Martin Appel, Judith Lessler, James Smith, William Eddy* (moderator)

3:00	Break
3:30-5:00	**Panel 4: Creating Statistical Information Products** *Michael Levi, Bruce Petrie, Diane Schiano, Susan Dumais (moderator)*
6:00-7:30	Reception
5:30-8:00	**Exhibits** TIGER Mapping System, Mable/Geocorr; U.S. Gazetteer; Census FERRET; CDC Wonder; National Center for Health Statistics Mortality Mapping Exhibit, Display, and Demo; Westat Blaise; Consumer Price Index CAPI; Census CAPI; FedStats

Wednesday, February 10, 1999

7:30 a.m.	Continental Breakfast
8:30	**Keynote Address** *Katherine Wallman*, Office of Management and Budget
9:00	**Panel 5: The Consumer's Perspective** *Virginia deWolf, Latanya Sweeney, Paul Overberg, Michael Nelson (moderator)*
10:30	Break
10:45	**Breakout Sessions** 1. Data management, survey technique, process, systems architecture, metadata, interoperation 2. Data mining, inference, privacy, aggregation and sharing, metadata, security 3. Human-computer interaction, privacy, dissemination, literacy
11:45	**Report Back from Breakout Sessions**
12:15 p.m.	Adjourn

PARTICIPANTS

RICHARD ALLEN, U.S. Department of Agriculture, National Agricultural Statistics Service
MARTIN APPEL, Census Bureau
DON BAY, U.S. Department of Agriculture, National Agricultural Statistics Service
LINDA BEAN, National Center for Health Statistics
LEWIS BERMAN, National Center for Health Statistics
TORA BICKSON, RAND Corporation
LARRY BRANDT, National Science Foundation
CAVAN CAPPS, Census Bureau

LYNDA CARLSON, Energy Information Agency
DAN CARR, George Mason University
WILLIAM CODY, IBM Almaden
EILEEN COLLINS, National Science Foundation
FREDERICK CONRAD, Bureau of Labor Statistics
HEATHER CONTRINO, Bureau of Transportation Statistics
ROBERT CREECY, Census Bureau
W. BRUCE CROFT, University of Massachusetts at Amherst
MARSHALL DEBERRY, Bureau of Justice Statistics
DAVID DeWITT, University of Wisconsin at Madison
VIRGINIA deWOLF, Office of Management and Budget
CATHRYN DIPPO, Bureau of Labor Statistics
SUSAN DUMAIS, Microsoft Research
WILLIAM EDDY, Carnegie Mellon University
JEAN FOX, Bureau of Labor Statistics
JOHN GAWALT, National Science Foundation
JIM GENTLE, George Mason University
VALERIE GREGG, National Science Foundation
JANE GRIFFITH, Congressional Research Service
EVE GRUNTFEST, University of Colorado at Colorado Springs
CAROL HOUSE, U.S. Department of Agriculture, National Agricultural Statistics Service
SALLY HOWE, National Coordination Office for Computing, Information, and Communications
TERRENCE IRELAND, Consultant
THOMAS KALIL, National Economic Council
DAVID KEHRLEIN, Governor's Office of Emergency Services, State of California
SALLIE KELLER-McNULTY, Los Alamos National Laboratory
NANCY KIRKENDALL, Office of Management and Budget
BILL LAROCQUE, National Center for Education Statistics, Department of Education
FRANK LEE, Census Bureau
JUDITH LESSLER, Research Triangle Institute
MICHAEL LEVI, Bureau of Labor Statistics
ROBYN LEVINE, Congressional Research Service
DENISE LEWIS, Census Bureau
GARY MARCHIONINI, University of North Carolina
PATRICE McDERMOTT, OMB Watch
TOM M. MITCHELL, Carnegie Mellon University
SALLY MORTON, RAND Corporation
KRISH NAMBOODIRI, National Coordination Office for Computing, Information, and Communications

MICHAEL R. NELSON, IBM
CLIFFORD NEUMAN, Information Sciences Institute, University of Southern California
JANET NORWOOD, Former Commissioner, U.S. Bureau of Labor Statistics
SARAH NUSSAR, Iowa State University
LEON OSTERWEIL, University of Massachusetts at Amherst
PAUL OVERBERG, USA Today
BRUCE PETRIE, Statistics Canada
LINDA PICKLE, National Center for Health Statistics
JOSEPH ROSE, Department of Education
CHARLIE ROTHWELL, National Center for Health Statistics
ALAN SAALFELD, Ohio State University
RAVI S. SANDHU, George Mason University
WILLIAM SCHERLIS, Carnegie Mellon University
DIANE SCHIANO, Interval Research
PAULA SCHNEIDER, Census Bureau
JAMES SMITH, Westat
KAREN SOLLINS, National Science Foundation
EDWARD J. SPAR, Council of Professional Associations on Federal Statistics
PETER STEGEHUIS, Westat
LATANYA SWEENEY, Carnegie Mellon University
RACHEL TAYLOR, Census Bureau
NANCY VAN DERVEER, Census Bureau
KATHERINE WALLMAN, Office of Management and Budget
LINDA WASHINGTON, National Center for Health Statistics
ANDY WHITE, National Research Council